Electronic Communication: Technology and Impacts

AAAS Selected Symposia Series

Routledge
Taylor & Francis Group

LONDON AND NEW YORK

for the

 American Association for the Advancement of Science
1776 Massachusetts Avenue, N.W., Washington, D.C.

Electronic Communication: Technology and Impacts

Edited by
Madeline M. Henderson
and Marcia J. MacNaughton

AAAS Selected Symposium **52**

First published 1980 by Westview Press, Inc.

Published 2018 by Routledge
52 Vanderbilt Avenue, New York, NY 10017
2 Park Square, Milton Park, Abingdon, Oxon OX14 4RN

Routledge is an imprint of the Taylor & Francis Group, an informa business

Library of Congress Cataloging in Publication Data
Main entry under title:
Electronic communication.
    (AAAS selected symposium ; 52)
    Based on a symposium which was held at the 1979 AAAS national
annual meeting in Houston, Tex., Jan. 3-8.
    1. Telecommunication. 2. Information networks. 3. Computer
networks. 4. Information storage and retrieval systems. I. Henderson, Madeline M. II. MacNaughton, Marcia J. III. American Association for the Advancement of Science. IV. Series: American Association for the Advancement of Science. AAAS selected symposium ; 52.
TK5101.E413   621.38   80-12431
ISBN 13: 978-0-367-02201-3 (hbk)
ISBN 13: 978-0-367-17188-9 (pbk)

About the Book

Electronic communications technology and services permeate every aspect of national life. This book examines the current and expected states of the technology and considers the societal impact and policy issues arising from new technological developments. Particular attention is paid to evaluation of computerized conferencing for enhanced communication among researchers in specialized and interdisciplinary fields and to technology assessments of criminal justice and tax administration systems.

About the Series

The *AAAS Selected Symposia Series* was begun in 1977 to provide a means for more permanently recording and more widely disseminating some of the valuable material which is discussed at the AAAS Annual National Meetings. The volumes in this *Series* are based on symposia held at the Meetings which address topics of current and continuing significance, both within and among the sciences, and in the areas in which science and technology impact on public policy. The *Series* format is designed to provide for rapid dissemination of information, so the papers are not typeset but are reproduced directly from the camera-copy submitted by the authors. The papers are organized and edited by the symposium arrangers who then become the editors of the various volumes. Most papers published in this *Series* are original contributions which have not been previously published, although in some cases additional papers from other sources have been added by an editor to provide a more comprehensive view of a particular topic. Symposia may be reports of new research or reviews of established work, particularly work of an interdisciplinary nature, since the AAAS Annual Meetings typically embrace the full range of the sciences and their societal implications.

WILLIAM D. CAREY
Executive Officer
American Association for
the Advancement of Science

Contents

Figures and Tables

About the Editors and Authors

Madeline M. Henderson *is the recently retired manager of ADP Information Analysis, Institute for Computer Sciences and Technology, National Bureau of Standards. She is a member of advisory committees on library networking and copyright law, chairman of the Joint Board/Council Committee on Copyright of the American Chemical Society, and author of a bimonthly column on copyright for the* Bulletin of the American Society for Information Science. *She has published numerous articles on information science and technology and library automation.*

Marcia J. MacNaughton *is a senior analyst at the Office of Technology Assessment, U.S. Congress, and has been a consultant to Congress on social impacts of computers and communication technology. She was staff director of the U.S. Senate Government Operations Subcommittee on Privacy and Information Systems and served on the staff of the U.S. Senate Judiciary Subcommittee on Constitutional Rights. Her primary interests are public law and government and information technology.*

John Bregenzer *is an associate professor in the Department of Sociology and Anthropology at the University of Dayton. His area of specialization is cultural change, and he is the author of* Trying to Make It: Human Adaptation to an Exposed Island *(Schenkman Publishing Co., in press).*

A. E. Cawkell *is vice-president for research, Department of Research and Development, at the Institute for Scientific Information, England. His specialty is information science, and he has written a number of articles, including "The New Technology in Information Science,"* Aslib Proceedings, *1979.*

Ira W. Cotton *is manager of social networking and office systems, Institute for Computer Sciences and Technology, National Bureau of Standards. He is a member of the editorial boards of* Computer Networks *and* Computers and Graphics *and has written numerous articles on data communications and computer networks.*

Linton C. Freeman *is professor and dean of the School of Social Sciences at the University of California, Irvine. He is the editor of* Social Networks *and has written 16 books and numerous articles on social network analysis, including* Communication Systems and Resources in the Behavioral Sciences *(with D. Easton, W. E. Schram and T. A. Sebeok; National Academy of Sciences, 1967),* Residential, Segregation Patterns *(with M. H. Sunshine; Schenkman, 1970), and* Bibliography on Social Networks *(Council of Planning Librarians, Exchange Bibliographies, 1976).*

Sue C. Freeman *is an evaluator for the National Science Foundation project on social networks. She is in the Department of Social Relations, Lehigh University, and is primarily concerned with the study of social structure.*

J. F. Hessman *is executive secretary of the Joint Electron Device Engineering Council (JEDEC) of the Electron Industries Association. He has specialized in electronic engineering, engineering standards and association management, and was principal investigator for the National Science Foundation Computer Conferencing Project.*

Starr Roxanne Hiltz *is professor of sociology and anthropology at Upsala College and associate director of the Computerized Conferencing and Communications Center, New Jersey Institute of Technology. She is also currently president of the New Jersey Sociological Society and a member of the advisory board of the International Network for Social Network Analysis. Among her numerous publications is* The Network Nation: Human Communication Via Computer *(with M. Turoff; Addison Wesley, 1978), which received the outstanding technical publication award (1978) from the Association of American Publishers.*

Walter R. Hinchman, *president of Walter Hinchman Associates, is a specialist in policy and systems analysis. He is a former chief of the Common Carrier Bureau of the Federal Communications Commission and a former assistant director of the Office of Telecommunications Policy, Executive Office of the President. He has published widely on telecommunications system engineering and policy.*

Peter and Trudy Johnson-Lenz *are independent consultants in computer communications applications. They work with clients both in the United States and abroad via Electronic Information Exchange System (EIES) in their home, and they exemplify in their own activities the shift from travel to telecommunications. They developed some of the first tailored applications subsystems within EIES, were coauthors (with M. Turoff) of the EIES users manual, and have carried out a variety of other projects on data analysis, design and development of software, and editorial processing entirely on line via EIES.*

Kenneth C. Laudon *is a professor of sociology at John Jay College, City University of New York. His areas of specialization are information policy, the social and organizational impact of computers, and statistics. He has written several books, including* Complexity in Large Federal Databanks *(Society/Transaction, 1980),* Telecommunications and Democratic Participation *(Praeger, 1977), and* Computers and Bureaucratic Reform *(Wiley, 1974).*

Joseph P. Martino *is a research scientist of the Research Institute, University of Dayton, and a specialist in technological forecasting. He is a fellow of the Institute of Electrical and Electronics Engineers and associate editor of* Technological Forecasting and Social Change. *He has published numerous articles on technological assessment and is the author of* Technological Forecasting for Decisionmaking *(American Elsevier, 1972).*

Jane H. McCarroll *is vice-president of Innovative Systems Research, Inc., and a consultant to various public agencies in special education research design and conceptual information system design. She was principal investigator for a National Science Foundation project on devices for the disabled and has worked on applications of computerized conferencing in the rehabilitation field.*

Carl H. Savit *is senior vice-president, technology, at Western Geophysical Company and is a member of the National Research Council-National Academy of Sciences Assembly of Mathematical and Physical Sciences and Energy Research Advisory Board, as well as chairman of the Committee on Geothermal Energy. A specialist in exploration geophysics and seismology, he received the Kauffman Gold Medal (1979). He has published many papers on and holds numerous U.S. and foreign patents in exploration geophysics and data processing.*

Elliot R. Siegel *is a specialist in scientific communication and information processing at the Lister Hill National Center for Biomedical Communications, National Library of Medicine.*

Murray Turoff *is director of the Computerized Conferencing and Communications Center of the New Jersey Institute of Technology. He has been a pioneer in developing the Delphi techniques and computerized conferencing. Among his many publications are* The Delphi Method: Techniques and Applications *(H. Linstone, coeditor; Addison Wesley, 1975) and* The Network Nation: Human Communication via Computer *(S. R. Hiltz, coauthor; Addison Wesley, 1978).*

Stuart A. Umpleby *is an associate professor of management science at George Washington University. His research has been concerned with cybernetics, computer conferencing, and applications of General Systems Theory and he has published a number of articles on these topics and on Delphi projections and alternative futures.*

Alan F. Westin *is professor of public law and government at Columbia University and president of the Educational Fund for Individual Rights. During the past 25 years he has written extensively about the impact of technology and social change on organizations, citizens, and society, particularly the issues of individual privacy and civil liberty vis a vis organizational and governmental record keeping and computerized information technology. He has served as consultant to business and state and federal governments on these issues and was principal consultant to the Senate Committee on Government Operations which drafted the Federal Privacy Act of 1974. Among his many publications are* Privacy and Freedom *(Atheneum, 1967),* Information Technology in a Democracy *(Harvard University Press, 1971), and* Databanks in a Free Society *(Quadrangle, 1972).*

Madeline M. Henderson

Introduction

Recent years have seen a revolution in information production and processing in our national economy. The complexity of our society has increased demands for timely and relevant information for management of organizations and institutions, for provision of education and entertainment, for conduct of governmental and social services. These demands are
being met, in large part, by innovative technological developments in computers and electronic communications.

Computers are of course an integral part of everyday life
in this country: computers write our paychecks and benefit
checks, then speed their use through electronic funds transfer
systems, automated checkout counters at the supermarket and
point-of-sale terminals in retail outlets. Computers provide
access to large data bases of specialized and general information for researchers and legislators, law-enforcement offi-
cials and laboratory scientists, students and administrators
alike.

But computers alone can be isolated; electronic communications capabilities make a vast difference. Communications
links put us in instant touch with colleagues around the country and with services at widely-scattered locations. The
electronic funds transfer systems just mentioned link banks
and clearinghouses across the nation; checkout counters and
point-of-sale terminals for chains and federations of stores
transmit data for centralized inventory and management control; specialized data banks, as in the law enforcement area,
extend the capabilities of local facilities to nation-wide
scope; large information files are coupled in library and information networks and made available through interactive terminals in home or office.

But decentralized operations and increased information transmission or communication can cause problems of their own - problems which include protection of the data being transmitted, allocation of available transmission channels, training and effective use of people. We can project less face-to-face communication even as the technology enhances overall communication via computer terminals, home computers, satellite transmission, and similar developments.

Moreover, there is potential for abuse in the ever-increasing use of computers coupled with electronic communications - privacy can be invaded, for example. And we can see increased regulation as a requirement for the burgeoning field and its industry. Large-scale national, even international, computer-based information systems will be - are being! - used to make very important decisions, sometimes without clear indication of the validity of the original data in the systems. Also, such facilities are more easily available to those with requisite resources - the rich and powerful organizations in our country have the advantage of increased information and communication systems, which in turn enhance power.

Recognizing these various components of a national resource and a national concern, members of Section T (Information, Computing and Communications) of the American Association for the Advancement of Science organized a symposium on the subject for the annual meeting of the AAAS in Houston, Texas on January 6, 1979. The symposium, entitled the Future of Electronic Communications, was co-sponsored by the American Society of Information Science, The Society of Exploration Geophysicists, and AAAS Sections E (Geology and Geography) and M (Engineering).

The symposium examined the current and expected states of the technology of electronic communications, as well as the societal impacts and policy issues arising from these technological developments. The electronic communications technology and services considered include interactive computer-based information and data systems, systems for handling large masses of data, electronic mail and message services, and systems linking computers, telephones and television in home or office.

In particular, the symposium and this volume started from a baseline overview of the current status and trends in electronic communications. Technological improvements in data transmission and message switching are reviewed, and a brief overview given of the policy and regulatory decisions which have been made or are yet to be made. The beginner in telecommunications use is given useful hints on getting started.

This overview is followed by examples of current and emerging applications of the technologies of computers and communications. In particular, developments in transmission and reduction of the large amounts of data generated in geophysical exploration for energy resources are recounted. And at the other end of the spectrum, so to speak, experiments with transmission and reception of science news on home television sets are described. Such experiments, underway in England for over a year, are just getting started in the United States and Canada.

Another process for electronic information transfer is computerized conferencing. Geographically-dispersed individuals can interact via computer terminals and communication links; each person enters material and/or comments into the computer store and retrieves others' comments and material as convenient. In the symposium, a panel was convened for the purpose of discussing the impact of computerized conferencing on the communication patterns and productivity of two specialized research communities. Panel members included representatives of the research groups, the designer of the Electronic Information Exchange System (EIES) which the research groups were trying, and the principal investigator of a study, sponsored by the National Science Foundation, of impacts of electronic information exchange, during operational trials, on a number of specialized research communities. The two communities participating in the panel discussion at the symposium had more experience to share with the symposium attendees than other communities participating in the EIES trials. In this volume, though, it has been possible to include information from most of the participants, representing a variety of research groups, as well as from the designer of EIES and the principal investigator of the evaluation study. The EIES is one example of a computerized conferencing and electronic information exchange facility; the lessons to be learned from its experimental use will be valuable to other such systems and their users.

Having reviewed the technologies of electronic communications and their various applications, the symposium and this volume turned attention to the societal impacts and policy implications attendant upon those applications. It is obvious to many of us that the technologies do not solve all problems and in fact cause several. The latter include, for example, questions of economic growth and distribution of benefits so that services will be universally available rather than developed only for the very rich. The issue as to whether competitive services should be encouraged so as to lead to lower prices and to stimulate productivity gains is debated. But policy dilemmas in this area are compounded by the relation

between such competitive services and the existing monopoly
services, e.g. telephone services: lower prices in the one
could result in higher prices in the other. National leadership is needed to tackle such matters.

Congressional concerns with these and other policy issues
are reflected in the number and variety of committees involved
with questions relating to electronic communications. The
then-current identity of such committees and their particular
concerns were enumerated at the symposium; however, since the
pattern changes rapidly and frequently, it was deemed inadvis-
able to attempt an updated version of that discussion for
this volume. But the general problems those Congressional
groups wrestle with are discussed, as they were at the symposium, in the last papers of the volume: accountability for
the content and management of large data bases containing personal data, the effects on organizations and social institutions caused by introduction of computers and electronic communications technologies, and the roles and responsibilities
of various governmental agents in this area.

This volume, and the symposium on which it was based,
recognize that the technologies of computers and electronic
communications are still developing, and that the problems
which follow their use are not solved. Rather, the presentations here constitute a snapshot of the current state and
discernible projected trends of the technology, and the present efforts to cope with the attendant problems. Those of us
involved with the symposium and this volume feel that this
picture will be a useful part of the total panorama of developments in the field.

Thanks are due to those who made the original presentations in Houston in January, and then updated or edited their
material for inclusion in this volume. We express particular
appreciation to Starr Roxanne Hiltz, who arranged for the
greatly-expanded discussion of experiences with computerized
conferencing. Special thanks go to Marcia MacNaughton who
played all the roles of symposium co-arranger, speaker, author, and volume co-editor. We thank Elaine Begeman who
transcribed most of the symposium recording. And we give an
extra-special thanks to Mary Ellen Crane, whose patience and
skill in the production of the individual chapters made this
volume possible.

Part 1

Current and Future States of Electronic Communications Technology

1. Perspective on Trends in Electronic Communication

\;

<u>Introduction</u>

This review and perspective will concentrate on the area
of electronic data communications, where major technological
advances have been made in the past ten or fifteen years. Not
only technological changes but relevant regulatory concerns
are of interest to us, and will be covered here. Their impact
on users will also be summarized.

<u>Communications Channels</u>

The history of technology in data communications has been
based on the premise that bandwidth is expensive and the efforts of technology are to conserve bandwidth, to pack more
traffic into lines of a given capacity. The first approach to
sharing the capacity of a communications channel was to divide
the channel into a fixed number of subchannels that could each
be used by a single user. (Figure 1). That's known as fixed-
division multiplexing, where the capacity of the channel is
subdivided, generally with fixed logic or "hardwired," such
that the total capacity of the carrier is approximately the
sum of the capacities of the individual subchannels.

In actual use, when people interact with computer systems
from interactive terminals, more often than not they are
scratching their heads and thinking about what they are going
to do rather than actually doing anything, so the traffic
through these channels is intermittent or "bursty." We'd like

_____

Ed. Note: Dr. Cotton's presentation was based on a set of
excellent color slides, which cannot be adequately reproduced
in this volume, with an accompanying commentary. The tran-
scription of the commentary and reproduction of several key
slides are given here.

- **FIXED CAPACITY DIVISION OF AVAILABLE BANDWIDTH**
- **GENERALLY HARDWIRED**
- **TOTAL CAPACITY=Σ INDIVIDUAL SUBCHANNELS**
- **UNUSED BANDWIDTH IN SUBCHANNELS IS LOST**

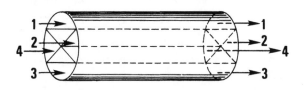

Figure 1. Multiplexing

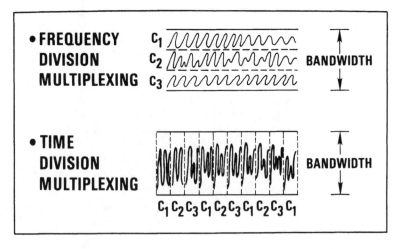

Figure 2. Multiplexing Techniques

to be able to recover some portion of this idle time and put
it to good use. But since each of those individual subchannels is unique and segregated from the others, when a user is
idle on a subchannel that capacity is irrevocably lost.

There are two fundamental approaches that are used to implement this kind of fixed-division multiplexing. (Figure 2).
In frequency-division multiplexing, which was just described,
the bandwidth of the channel is sliced up into smaller bandwidth sections, each of which can be transmitting continuously. The other approach is more like what is done in computer
time-sharing, where everybody gets the full capacity of the
channel but for a limited period of time; that is, one user
has the full bandwidth for a brief period of time, then the
next user, and the next, then back to the first user again.
This is called time-division multiplexing. But the situation
still holds that if a user doesn't have anything to send at a
particular point in time, that entire time slice stays empty,
is unused and is lost. In addition, in this method the identification of data is implicit in the order in which they are
sent; that is, every n-th chunk of data will belong to a particular user. In order to recapture unused capacity by skip-
ping a user, some identification of whose data are being
transported will have to be carried along with the data.

That is precisely what is done in concentration, also
called adaptive (or statistical) multiplexing, where the channel capacity is divided in a variable rather than in a fixed
way. (Figure 3). Under program control, with some intelli-
gence in the communications system, the bandwidth is divided
into continuously varying pieces or allocations. In this
case, there has to be some way of knowing when a new portion
of data begins and when it ends; in other words, we must be
able to identify to whom the data belong. This introduces
some degree of overhead, so that the total capacity of the
channel is somewhat less than the information-carrying capacity of the logical subchannels. On the other hand, if a
group of users is sharing a channel in this way, if any of
them has anything to send there won't be any idle capacity,
the channel will be occupied and there can be much more efficient utilization of the capacity.

When we compare the multiplexing and concentration approaches to sharing bandwidth, we see that the theoretical capacity of the multiplexer is really the full channel capacity
because there is no overhead carried. If there is continuous
traffic from a number of sources where the peak-to-average
ratio is very low, that is, there is quite a regular flow of
information, then a multiplexer is the most efficient way to
share traffic. With concentration, the capacity of the carrier is reduced by the amount of overhead that is introduced,

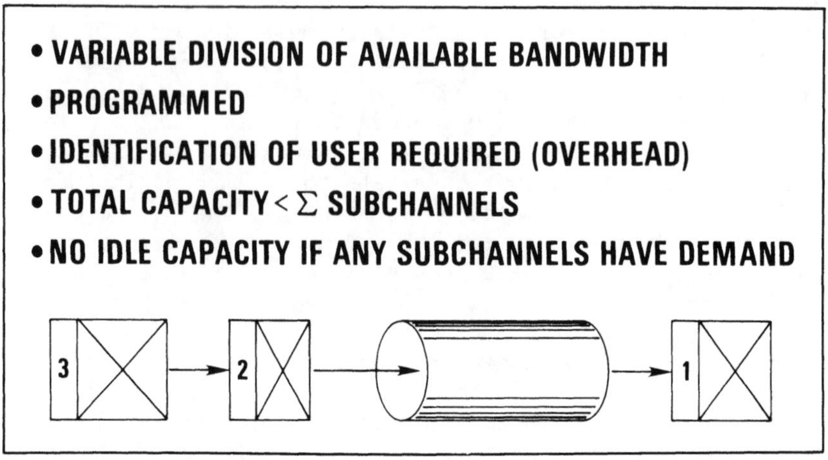

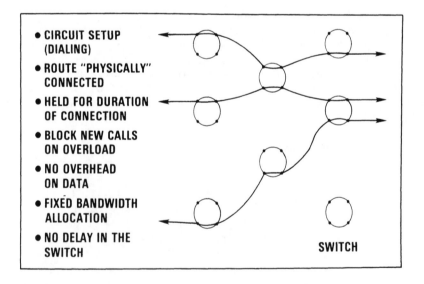

Figure 3.   Concentration: Adaptive (Statistical) Multiplexing

Figure 4.   Physical Circuit Switching

so theoretically it is not as efficient as the multiplexer.
What happens in actual fact is that the capacity of the multiplexer in operation is considerably reduced by any idle time
from any of the transmitters. In the concentrator, only the
overhead is lost and all of the idle capacity is regained, so
that in most cases a concentrator is much more efficient than
a multiplexer, particularly for interactive terminals, where
the peak-to-average ratio is typically high.

Switching

   Switching is the process of selecting a path for information to go from sender to a number of alternative receivers.
In talking about physical circuit switching, our model is the
voice telephone network in this country. (Figure 4). A simplified model of the phone system shows that when a number is
dialed there is first a circuit setup phase, a path through
the network is found, and that route is in some sense physically connected. (In the old days, a network of operators
actually plugged in the long distance calls!) This circuit or
path is fixed and is held for the duration of the call. New
calls are blocked on overload; that is, when the system approaches saturation it refuses to accept new traffic. There
is no overhead on data, there is a fixed bandwidth allocation
and there is no delay in the switch. Those are the salient
characteristics of this kind of switching technology.

   What is done in the way of physical connecting of the
switches in setting up a fixed path, can also be done with a
minicomputer. (Figure 5). Rather than having the input line
and the output line physically connected, they can be connected logically or virtually with a program in a minicomputer. A
network in which the physical switches are replaced with
stored program devices still has a circuit setup phase, then
the route through the network is logically connected by tables
in memory, rather than being physically connected by plug
boards or cross-box relays. New calls are still blocked to
prevent overload. One difference in this approach is there is
probably some overhead on data in order to identify to whom
the data belong. Bandwidth can be allocated when the call is
placed, because the amount of time allocated to a user is a
function of the program in the minicomputer, and different
amounts of bandwidth can be allocated.

   Another significant difference between physical circuit
switching and virtual circuit switching is that in the latter
case data are stored in the switch: a block of data comes in,
it is stored in memory while a decision is made on how to handle it and then it is re-transmitted. This stored delay is
cumulative so for longer distances the data are increasingly
delayed in transmission from sender to recipient.

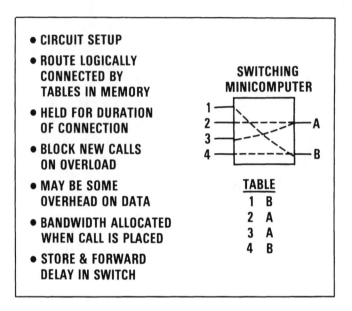

- CIRCUIT SETUP
- ROUTE LOGICALLY CONNECTED BY TABLES IN MEMORY
- HELD FOR DURATION OF CONNECTION
- BLOCK NEW CALLS ON OVERLOAD
- MAY BE SOME OVERHEAD ON DATA
- BANDWIDTH ALLOCATED WHEN CALL IS PLACED
- STORE & FORWARD DELAY IN SWITCH

SWITCHING MINICOMPUTER

TABLE	
1	B
2	A
3	A
4	B

Figure 5.   Virtual Circuit Switching

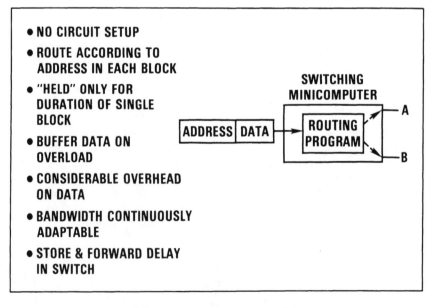

- NO CIRCUIT SETUP
- ROUTE ACCORDING TO ADDRESS IN EACH BLOCK
- "HELD" ONLY FOR DURATION OF SINGLE BLOCK
- BUFFER DATA ON OVERLOAD
- CONSIDERABLE OVERHEAD ON DATA
- BANDWIDTH CONTINUOUSLY ADAPTABLE
- STORE & FORWARD DELAY IN SWITCH

SWITCHING MINICOMPUTER

ADDRESS | DATA → ROUTING PROGRAM → A / B

Figure 6.   Message Switching

Given a minicomputer in place, one is not constrained to have the data follow the same path all the time, so we can progress to what is called message switching. (Figure 6). Here the circuit setup is abolished, and data can be routed through this network on a block-by-block basis according to the address. There's no holding time in this kind of a system, that is, there is no fixed route that's reserved for an individual user that would serve to block other users. When the system becomes overloaded it simply stores the data in memory; service will gradually degrade to all users as the system becomes more and more heavily used. There may be considerable overhead due to the addressing information and the functioning of the routing program, but the bandwidth is continuously adaptable. So long as the system is not saturated, users get as much or as little bandwidth as they need and the system can adapt to selective component failure. If a switching center or a circuit fails, data can be sent through alternate routes quite easily to the ultimate destination.

Packet switching is a subset of message switching with a number of characteristics, no single one of which is sufficient to completely discriminate between the two, but which when taken together describe a quite different kind of service. First of all, the block length or the transmission slice is kept relatively small, perhaps no more than a hundred characters, for example. The switch is designed for forwarding, not for storing: people talk about "hot potato routing," wherein a packet of data comes in and has to be either passed on immediately or thrown away. There is no backup storage, no extensive records of traffic, and again, input is refused to avoid overload. When the system becomes saturated, information is simply thrown away, undeliverable packets are discarded, and it is up to the sender and the receiver to maintain their own control scheme to guarantee the receipt of information in its proper sequence.

Now a data communications network might be called upon to carry different kinds of traffic: interactive (transaction) traffic, with a user at a terminal, intermittent or bursty flow and, since users of terminals don't like to wait, the requirement for rapid response from the system. In contrast, bulk traffic or remote batch entry involves large jobs that may be transmitted at night, that have a continuous moderate-to-high bandwidth demand, and lessened response requirements so that bulk traffic is much easier to engineer for than interactive traffic. The final category of real-time traffic combines the most perverse characteristics of the first two: high bandwidth bulk traffic also requiring high response, the kind of traffic in command and control systems, in sensor applications and laboratory automation. Data that cannot be delivered in an adequate time period are probably useless for the application

Let's consider how these types of traffic can be handled
by circuit switching (either physical or virtual) or packet
switching. Interactive traffic is somewhat unsuited for circuit switching and that's really what led to the concentration
and multiplexing schemes described earlier. Without a well-
defined multiplexing scheme, there is going to be a lot of
wasted bandwidth. Packet switching can have potentially high
overhead but there is not going to be any idle bandwidth. It
turns out that packet switching is ideal for terminal-to-com-
puter traffic where people are occasionally transmitting messages through a computer system. Bulk traffic on the other
hand is most suitable for circuit switching; with a lot of
continuous information to send, utilization of the line is
quite good. The suitability of bulk traffic for packet
switching is really a function of price. With current pricing, it is most cost effective for moderate quantities over
long distances; over short distances you are better off dial-
ing a call. Over long distances or with lots of information,
you are better off leasing a line. Real-time traffic is very
suitable in the circuit switching case with dedicated facilities: a leased line, guaranteed bandwidth and certainty that
the transmitted data will get through. There have been experiments in the research community to carry real-time traffic
over packet switching systems; it is only possible with some
new control mechanism added to the network to give certain
packets priority over others, to guarantee that they get
through in a reasonable period of time.

Regulatory Background

For data communications, one of the first significant
regulatory proceedings in this country came, back in the late
1960's and early 1970's, when the Federal Communications Commission (FCC) considered the issue of specialized common carriers. The question on the FCC docket asked specifically
whether carriers other than the established ones could offer
competitive services. It took eight years for the FCC to finally set the course for free competitive entry into selected
areas of the communications marketplace. MCI was permitted to
erect its own microwave transmitter between Chicago and
St. Louis; based on this decision, there have been a substantial number of new entrants offering different kinds of service in the communications area in this country.

Another significant issue is the interdependence of
computers and communications. Here the question is, what do
you regulate? The Communications Act of 1934 gives the FCC
the authority to regulate communications, but it does not include the authority to regulate computers. The FCC has recognized that there's a spectrum between pure computing and

pure communications. In pure computing, e.g., time-sharing
services, the use of communications is only incidental to the
primary function of computing. This function is unregulated.
Pure communications, the phone system for example, may require
computation and logic decisions, may contain minicomputers in
the central offices and so forth, but they are only incidental
to the primary service of communicating. This would be regulated. So the "incidental" test asks whether the one function
is only incidental to the other. In the case of message
switching, it seems clear that this is communications, even
though there are additional components of computation, and so
it is regulated. But a series of hybrid services exist for
which the incidental test fails. Also, regulated carriers
were not permitted to offer computation service unless through
an independent subsidiary. So we had a spectrum of regulation, with unregulated computing services at one end, regulated communication services at the other end, and a grey area
in the middle where it wasn't clear whether a service would be
regulated or not.

Another question related to resale: service from the common carrier cannot be resold to end users, although there were
certain tariffs that permitted groups of users to share the
use of a communications system (Tariff 260, Joint Use Agreement). Companies like Tymshare were setting up shared-user
networks based on this tariff and this led to the concept of
the value-added carrier, where additional services such as error correction, automatic routing and interfacing were added
to the basic service of communications. The whole thing came
to a head in 1973 when Packet Communications, Inc. (PCI) asked
the FCC to regulate the value-added network market by filing a
Section 214 application and asking to be a regulated carrier.
PCI was granted this approval but never got adequate funding
to be able to offer the service. Telenet the following year
got similar approval and has been in operation since about
1975. Tymnet has since filed and a number of other carriers
have indicated their intention to enter this value-added network marketplace.

Finally, the computer inquiry was recently reopened by
the FCC which proposed a definition of data processing: "The
use of a computer for processing information wherein (A) the
semantic content or meaning of input data is in any way transformed, or (B) where the output data constitute a programmed
response to input data." Again the FCC is trying to make
clearer what will be regulated and what will not be regulated.
Comments on that definition have been filed and the docket has
not been closed yet. So there are some open regulatory questions that users need to be aware of.

User Alternatives

A prospective user of data communications can build a network, can buy an existing network or a customized version of an existing network, or can connect to an existing network. There is a wide variety of networks, including research and development networks, e.g., ARPANET and various regional networks; commercial public networks, including value-added nets such as Tymnet and Telenet; and private networks for corporate or industry groups, which are increasingly shifting over to public networks. (I think it is important to point out that there have and continue to be failures in networking. PCI failed to get adequate backing to ever offer service, Datran failed commercially and the policy of free entry is probably going to continue to result in a policy of free exit also.)

The value-added network (VAN) services purchase basic services from common carriers and add capabilities to them, as noted earlier, to produce new kinds of service for users. These include speed conversion, code conversion, and common protocols. In addition to the management of the total multi-vendor system the VANs offer a kind of pay-as-you-go approach, where the user can pay on a volume-of-data basis rather than a period-of-time or a fixed-capacity basis; that is quite important to a number of users. In comparing value-added networks, a variety of factors have to be considered, including fixed installation charges and fixed monthly charges, as well as the variable charges which for most users will be dominant. That is, there are charges which, in contrast to leased-line service and phone system charges, are based on volume of information. They are typically independent of distance, which is the primary reason for long-distance users making use of them, because that is where the major savings accrue. In current systems based on the use of landlines, distance-independent charges are really a marketing strategy rather than something based on technology. But as these carriers shift to using satellites, it is no more expensive to transmit up to a satellite and down across the street than up and down across an ocean, so that distance-independent pricing is somewhat anticipatory of new technology.

If we make a chart comparing typical tariffs for a variety of different services, plotting dollars per month versus distance, we see that except for very short calls of limited duration it is ridiculous to use dial-up phone systems for any major data communication applications. Leased-lines charges are more or less linear, increasing with distance, and the value-added network charges are flat right across the graph. So it depends on the volume of information and the distance as to whether a leased line or the use of a VAN is more

economical. As the volume of information increases for the
value-added network curve, that flat line would move up on a
parallel line and as the volume decreased, it would move back
down again. That is the basis for my statement that it is the
moderate-volume users over long distances who will find the
use of value-added networks much more economical.

Getting Started in Networking

The range of actions for beginning to use networks runs
from somebody giving you a phone number and a log-in procedure
for your terminal, all the way down to designing and imple-
menting your own host-to-network interface, which may take a
year or more of expert system programmer work. The communications environment today reflects that the cost of communications has been coming down, but bandwidth is still expensive
relative to other components involved.

The value-added networks are serving as a marketplace for
a variety of services and we find ourselves in an environment
where users likely need to access many different systems, perhaps in the same day. This leads us to the problem of learning multiple languages, remembering the log-in procedure for
one computer and trying to use it on another, talking French
to your German friends and vice versa. We see two alternative
solutions here. First is standardization. Being from the
National Bureau of Standards, I naturally see that as an ultimate solution but I also recognize that it is not going to
happen soon. There is just too large an investment in host
operating systems, software, and job control languages to hope
for an early acceptance of new standards in this area. So we
are investigating the second approach, of automatic transla-
tion, where we can take something we call a network access
machine, or a minicomputer, or a microprocessor in a terminal,
and translate on behalf of the user from a common command
language into the target language for a number of different
systems. At NBS we have implemented on a test basis a common
command language for five bibliographic retrieval languages,
we have it operating and are able to access with this common
language a wide variety of systems such as Lockheed, SDC and
others with quite different underlying syntaxes. The syntaxes
are significantly different and we are able to mask these differences through the use of our minicomputer system. Future
systems will probably include processors within terminals and
then we can provide such network access facilities at low
cost.

The key research issues that remain at the present time
include the integration of the various switching technologies
we have described. I think voice and data will be integrated

in the sense that voice will increasingly be carried digital-
ly. There is going to be for the remainder of the century an
increased emphasis on digital switching and carrying all kinds
of analog data in digital form. Local area networks are becoming increasingly important and anything that can be done to
cut costs there will be a major advance. Protocols and the
inter-connection of different types of computing systems
through networks, the general problem of distributed processing, network operating systems, and user access systems such
as I have just discussed, are all important research issues.

Conclusions

If you build your own network, it will be cheaper but it
will require somewhat higher investment and the need for continuous management, which is no small consideration. If you
have someone build it for you, you avoid investment in a permanent design staff, but you generally need to have a larger
net to justify this method. The big advantage of value-added
networks is that they are lowering the cost of entry. You can
connect to them quickly, easily, and pay only for the service
that is used. The cautionary note is that high demand users
may find these services more expensive.

2. A Quadrillion Geophysical Data Bits Per Year

Introduction

This year the geophysical exploration industry will record and use about one quadrillion (10^{15}) bits of geophysical data in its world-wide quest for more and better knowledge of the structure and content of the earth beneath our feet. All the significant information contained in this incredible mass of data must somehow be communicated to the few thousands of people that use this information in their daily work.

Over the past decade or so, the geophysical profession has been evolving a communications chain that effectively moves the geophysical data from its points of acquisition through several communication links and compaction steps and finally to the user community. New technological developments in the geophysical part of the system are, however, producing data more rapidly than the advancing state of the arts of communication can accommodate.

Communication from the Sensors

Nearly all of the high-volume geophysical data is derived from reflection seismology. In this method, hundreds or thousands of sensitive motion detectors are deployed on the earth's surface or are towed near the surface of the ocean. The detectors produce electrical voltages as they respond to the earth motion initiated by a powerful sound source. Until quite recently all of the electrical signals from the detectors travelled down telephone-type wires to a set of instruments in a truck or a boat and were then converted into numerical values to be recorded on magnetic tape.

Ed. Note: Mr. Savit's presentation consisted of a number of excellent color slides which cannot be adequately reproduced in this volume. His commentary is summarized here.

Because of major increases in the number of detectors being used, the number of "telephone" wires needed has begun to get out of hand.

Some of the newer geophysical field systems are being built with all of the data-conditioning electronics out in the field at or near the detectors. In such systems numerical values instead of voltages have to be communicated to the central station in the field. Such number values can be communicated down a coaxial cable telemetry line with almost perfect reliability. One system presently operating transmits some 600 separate signals down one coaxial line (actually, for extra reliability, 3 lines are used, each carrying the same data). To use a pair of telephone lines for each of the signals would have required the impossible number of 1200 wires in a portable cable. The technology that makes possible the sub-miniaturized electronics needed to make this system work originated with the space program.

Communication to the Computer

After reaching the recording instruments in the field, geophysical data must somehow be transmitted to a large scale computer, there to begin the complex series of data-reduction steps needed to compress the raw data and to extract its useful information content.

Ideally we would prefer to transmit our field data directly to a computer center while it is being taken. The method most frequently proposed is the use of geo-stationary communications satellites to relay data from a field truck or ship to a central computer facility.

Intriguing as this possibility appears, it has merely led several successive generations of system designers to ultimate frustration. As each increment of communication-channel capacity has been announced by satellite operators, the geophysicists have contemporaneously introduced instruments with higher data-production rates. The net result has been and still is that each generation of satellites has been capable of handling only the previous generation of geophysical data.

Today's satellites are just beginning to have a maximum link capacity of about 250,000 bits per second, but the new 5 or 6 hundred channel seismic systems generate about 10,000,000 bits per second. Even the slowest of 1979 model systems would require transmission rates of 1,000,000 to 2,000,000 bits per second.

It does, however, appear that at present the satellite limitations are not technological but institutional in nature. There is some hope that competitive forces will eventually eradicate the present constraints.

In the meantime, geophysical data are recorded on magnetic tape and physically transported to computer centers. The amount of tape used for this purpose is enormous. My organization, Western Geophysical Company, by itself consumes more computer tape than does any other entity in the world except for the U. S. Government.

Within about two years one may expect that tape will be replaced by the laser-etched disks that are being developed for the video-recording industry.

Communication with the End User

In the first two communications links of the geophysical chain, the on-going data explosion can be, and is being, accommodated by technological means, albeit not without some cost and effort. On the other hand the third and final link, the link between the computers and the human end-user, appears to pose a much more severe problem.

In past years much of the increased data volume has been applied primarily to enhance the quality of final output data rather than to increase its quantity. Recent developments in geophysical exploration have, however, begun to produce more kinds of information in more detail, about more closely-spaced parts of the earth's crust. An immediate consequence is that the individual user has to see and absorb much more information before making a decision about a given portion of the earth.

Taking advantage of the human ability to absorb vast amounts of information in pictorial form, the computing machinery of the geophysical industry more than a quarter century ago started to output graphic cross-section representations of the earth. Progress during the intervening period consisted primarily of increasing the amount of detail in these presentations.

We have reached a virtual limit in the amount of data that can be presented for viewing or apprehending as a single entity. For most areas today, several cross-sectional representations are made of the same survey line to represent different types of information. Some of these sections augment the amount of information they must communicate by representing some of the results in the form of computer-generated colors. Finally, many cross-sections for closely spaced parallel or intersecting survey lines are presented to simulate a three-dimensional representation.

Unfortunately, no presently known or imagined technique will present to a human a representation of the entire interior of such a highly variegated solid body as a block of the earth's crust. All we can do is look at two-dimensional slices or layers one at a time.

It seems reasonable to me that the inherent communication limit that we face is governed by the three-dimensional nature of our world and our inability to step into a fourth dimension to look inside a solid. If we are indeed to make use of the technological possibilities of geophysical exploration, we will have to train our computers to do much of our work for us. In particular, we will have to delegate decision-making, judgment calls to our mechanical alter egos.

The last communication link will then only require that the computer advise us of its decision and we need only see to it that the decision is carried out.

3. Viewdata, Prestel, Teletext: Home TV and Scitel Science Magazine

Introduction

"A Memex is a desk comprising translucent screens on which may be projected book pages, records, photographs, memoranda, all sorts of things; associative indexing enables any item to be caused at will to select another immediately and automatically."

> Vannevar Bush
> "As we may think"
> The Atlantic Monthly, 1945.

"So it can be done. Will it be done? Ah that is another question ... A long time from now, I fear, will come the personal machine. It will be delayed in coming principally by costs, and we know that costs will go down, how much and how rapidly none can tell."

> Vannevar Bush
> "Memex Revisited"
> Science is Not Enough
> (Apollo, NY 1969)

The above quotations, by an eminent American, convey both hope and uncertainty - note particularly the uncertainty about costs. Today we can be rather sure that a machine something like the one in Bush's vision will soon be available at a reasonable cost. Surely, you will say, the inexpensive personal computer is here now? Indeed it is, but at least three ingredients of the Memex are missing. First, large files of information cannot yet be cheaply stored on a home computer. Second, information has to be indexed in some way if we are to find quickly what we need - and that is an intellectual task which remains a problem. Third, cathode ray tube screens are too small; the eye/brain can easily scan and

select from, say, a newspaper page containing 30,000 characters, whereas the best CRT can manage about 2,500 without room
for graphics. A television receiver can manage about 1,000.

Some of these difficulties are being solved in a way
which Bush could hardly have anticipated. I am referring to
a potential pseudo-Memex in most people's homes - a television
receiver, plus a dial telephone, plus a home computer; although this paper is about Viewdata/Prestel the future interdependence of these three devices must be taken into consideration, and I will discuss them later.

Equipment Developments

In Britain, textual information displays for home consumption on modified television receivers have been developed
over the last few years in the following ways.

In 1972 the British Broadcasting Corporation (BBC), the
Independent Television Authority, the TV receiver manufacturers and the British Post Office jointly discussed a standard
for the transmission and format of pages of text to be displayed on a modified colour TV receiver. Transmission would
be digital - that is characters would be transmitted as codes
of on-off pulses or "bits". After decoding, one page of 24
rows of 40 characters would be stored and continuously displayed on the receiver. Provision would also be made for the
display of graphics, built up as a pattern of stored picture
elements.

The receiver would be a standard colour television set
modified by the addition of decoding, control, and storage
modules formed from integrated circuit elements. The transmitted data would contain information about the position and
row of each character on a page or the position of graphic
elements, the colour in which a character could be displayed,
and various other enhancements such as flashing or special
characters. This standard was adopted so that organisations
intending to transmit information and receiver manufacturers
could enjoy a period of technical stability.

Development of systems then proceeded along two paths.
The first system was called Teletext, and the Ceefax and Oracle television services, using this system, are now operated
by the BBC and Independent Television Authority (ITA) re-
spectively.

In Britain, television entertainment fields are transmitted 50 times per second, and Teletext codes are included as
bursts of data fitted into a small unoccupied part of each

field. It takes about 1/4 second to transmit one page of
text, and a sequence of 100 pages is transmitted every 25
secs. A viewer sets a page number on a dial on the receiver
and then waits 12.5 secs. on average until the wanted page is
captured the next time it is transmitted.

A typical page from these Teletext services, photographed
from a receiver*, is shown in Figure 1. Both Ceefax and Oracle are now operating in Britain on a regular basis.

```
 ┌─────────────────────────────────────────────────────┐
 │                                                       │
 │        Oracle p 253 Mon 05 Jan  19.07/18              │
 │                                                       │
 │             ITN  SPORTS  NEWS                         │
 │                                                       │
 │        Scottish League - Division Two                 │
 │                                                       │
 │   Top Four:         P   W   D   L   F   A   PTS        │
 │        Clydebank   20  13   4   3  35   9   30         │
 │        Raith       19  10   9   0  35  19   29         │
 │                                                       │
 │   Bottom Four:                                        │
 │        Brechin     17   5   4   8  23  27   14         │
 │        Stenh'smuir 19   4   4  11  22  35   12         │
 │        Forfar      19   3   6  10  19  34   12         │
 │        Meadowbank  20   3   5  12  12  44   11         │
 │     League Division One follows shortly               │
 └─────────────────────────────────────────────────────┘
```

Figure 1

Meanwhile the second system for text display, named
Viewdata, was being developed by the British Post Office. A
service called Prestel is running experimentally in Britain
using this system; it is rather different from Ceefax and
Oracle. The same modified television receiver, codes, and
storage are used, but the receiver is connected to the viewer's telephone line and receives information by this route
instead of via the TV antenna. Additionally the viewer holds
a small keypad associated with the TV receiver embodying buttons marked 0 to 9, two control buttons, and an auto-dialing
button; with this device command signals for particular pages
may be sent along the same telephone line. The heart of the
system is a local computer which may be dialed by a viewer;
the machine now operating from the London area can handle 100
calls simultaneously, but will soon be able to handle several

───────────────────────────

* *Ed. Note:* In Mr. Cawkell's presentation, colored slides
showed pages photographed from the TV screen; typed representations are used here.

VIEWDATA/PRESTEL

Some of the services -

News	House & Mortages
Sport	Going Out
Radio & TV	Facts & Figures
Holidays & Travel	Money & Insurance
Tourist Guide	Jobs & Careers
Education	Business Services
Hobbies & Pastimes	Messages
Jokes, Quizzes, Games	Science & Technology News
Home & Family	Cars & Motoring
Marketplace	

Figure 2

Prestel   1791a   0.5p

INFORMATION NOW ON PRESTEL

1 BUSINESS ½p Shares, money, companies
2 SPORT & HOBBIES ½p Taking part, recds
3 ENTERTAINMENT ½p Places to visit,
    jokes, quizzes, eating out, theatre
4 HOLIDAYS, TRANSPORT, TRAVEL ½p Hotels,
    package tours, air, sea, rail, motoring
5 MARKETPLACE ½p For sale, consumer aid
6 ADVICE ½p Medical, educational, money,
    safety, buying advice, legal
7 GOVERNMENT INFORMATION ½p
8 EMPLOYMENT ½p jobs, careers, law
9 BOOKS & REFERENCE ½p facts, reviews
0 OTHER ½p Science, weather, charities,
    horoscopes

PRESTEL^{tm}

Figure 3

hundred. Similar machines will be installed in other parts of
the capital and in other cities during 1979. Upon receipt of
a command from a viewer, the computer retrieves page data from
storage and transmits it down the line to be displayed on the
TV receiver.

There are three basic differences between Viewdata and
Teletext. First, the viewer can select a Viewdata page, then
proceed to a second page in an interactive manner, in contrast
to the more passive selection procedure on Teletext. Second,
a dedicated channel - the telephone line - is available for
sending and receiving information; an ordinary switched telephone line has the capacity to transmit text codes with virtually no delays, in contrast to the Teletext system. Third,
the number of pages available is limited only by economic and
computer storage considerations. This is expected to be several hundred thousand per computer, with perhaps a larger number available from bigger machines networked into the system.
As of December 31, 1978, about 100,000 pages are stored in the
London Viewdata/Prestel machine.

The source of information in Teletext is the BBC and ITA.
On Viewdata/Prestel any organisation can provide a service by
arrangement with the Post Office. Currently about 150 different organisations have services operating at various stages
of development. The kinds of service which are becoming
available are shown in Figure 2. It will be obvious that if
only about one hundred pages are available on Teletext, then
they can be indexed rather easily. For instance, a particular
page can be set aside for a subject index to other pages. For
the larger number of pages on Viewdata a special index has
been devised rather like the branches of a tree. The viewer
is first presented with ten general choices of subjects numbered 0 to 9; upon pressing the appropriate button on the keypad ten numbered sub-divisions of the chosen subject are presented, and so on, until the information page is displayed.
It will be obvious that 1 out of 100,000 pages could be chosen
by 5 successive "choose 1-out-of-10" commands. If, however,
the viewer already knows the number of a wanted information
page he can send a direct command for it.

Scitel Services

The way Prestel works in terms of one of its component
services - Scitel - is shown in Figures 3 - 12. Figure 3 shows
Prestel's very first page.

If button '0' is selected the choices shown in Figure 4
are presented.

After two more selections along the science route the
first page of Scitel appears. The page shown in Figure 5 then
follows. Each page always ends with a prompt - usually either
a 'continue' instruction (the "#" - one of two special buttons
on the keypad) or a "key page number" instruction.

Next comes the Scitel index page (Figure 6), and it seems
appropriate at this point to discuss the content and purpose
of this service in the context of the total Prestel services.

```
Prestel                    17910a    0.5p

OTHER INFORMATION

1   SCIENCE 0.5p
2   WEATHER 0.5p
3   HOROSCOPES 0.5p
4   GIVING TO CHARITY o.5p
5   PARLIAMENT 0.5p


PRESTEL^tm
```

Figure 4

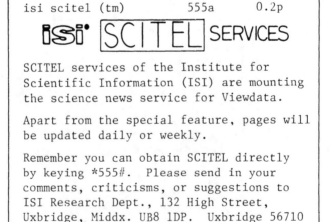

Figure 5

Prestel is currently running experimentally with up to
1500 receivers - half in businesses and half in households -
installed by the Post Office for market research purposes.
This phase will merge into a public service to be inaugurated
in various parts of the country during 1979. The business of
the Institute for Scientific Information(ISI) is self-evident,
and when thinking about getting into this new media we had to
consider what information we were able to supply and the interests and size of the developing audience. We chose to experiment with a science magazine - modest but capable of being
easily expanded. The service is aimed at informed laymen and
it attempts to provide an overview of current science. To
this end subject coverage and topicality have received careful
attention. Areas of science are grouped as shown in Figure 6:
we obtain our information from press releases, foreign embassy
communiques, government departments, research establishments,
scientific journals, newspapers, and so on. The information
is selected, edited, and highly compressed for input to the
Prestel computer on-line from a terminal in our research offices at Uxbridge. Information can be input immediately at
any time of day - a small fraction of what we receive demands
immediate treatment. Other material - such as book reviews -
can be handled in a more leisurely manner. Some information
becomes rapidly out of date and is quickly replaced; other
items may endure for several weeks. At present we have about
200 pages in central storage, and the rate of change is about
50 pages per week.

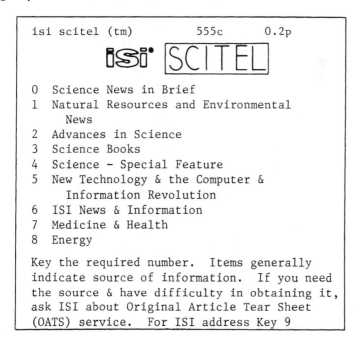

Figure 6

Continuing from Figure 6 - the index page - examples of information pages are shown in Figures 7 through 12. Note the facility for cross-referencing to a more detailed exposition from a terse announcement by a single button prompt. This technique may be used to fetch any page from the entire Prestel database.

```
isi scitel (tm)              55500b     0.2p

SCIENCE NEWS HEADLINES

Theme for the annual meeting of the American
Association for the Advancement of Science,
Houston, Texas, Jan 3-8, 1979, will be Science
& Technology: resources for our future
                For more - Key 4

Exploration of mud from bottom of Red Sea
indicates presence of silver, zinc and
other metals in workable quantities
                For more - Key 5

Dental care surgeries are being set up in
stores in California.  Retail dentistry "is
the coming thing" says the Los Angeles Times
quoted in The Guardian, 19.12.78.
           For more News Headlines - Key #
```

Figure 7

```
isi scitel (tm)              55504a     0.2p

SCIENCE NEWS:   POLICY

AAAS Meeting, Houston, Jan 3-8, 1979
Science & Technology - resources for
our future

This large meeting will comprise public
lectures, a science film pictorial, and a
symposium.  Amongst the subjects will be -

    *  Ecology & the Environment
    *  Social Sciences & the family
    *  Education
    *  Energy
    *  Engineering
    *  Science Policy
    *  Agriculture & Nutrition

The technology and .......
             To continue - Key #
```

Figure 8

```
isi scitel (tm)          55504b    0.2p

POLICY (CONTINUED)

AAAS Meeting

... consequences of the rapidly expanding
field of information & communication (in-
cluding Prestel/Viewdata) will also be
discussed.

The public lectures include such topics as -

   -  Space exploration
   -  Understanding arterio-sclerosis
   -  The seasons of men
         For more Policy News - Key #
```

Figure 9

```
isi scitel (tm)          5552a     0.2p

            ADVANCES IN SCIENCE
            ********************
            0   Life Sciences
            1   Physical Sciences
            2   Social Sciences

Key the required number
```

Figure 10

```
isi scitel (tm)          55522c    0.2p

ADVANCES IN SCIENCE:  SOCIAL STUDIES

U. S. population will peak by 2015

Population of the US is growing much more
slowly than was forecast only a few years
ago, and is likely to peak at 253m in 2015.

This is well below the forecast of 300m by
the year 2000 that had been widely used until
fairly recently.  The population of the US
is currently estimated to be about 220m.

Relevant factors are a diminishing number
of women who marry and the postponement
of marriage to a later age.  In 1960,
for example.....

To continue - Key #
```

Figure 11

```
┌─────────────────────────────────────────────────┐
│  isi scitel (tm)            55522d    0.2p        │
│  SOCIAL SCIENCES (CONTINUED)                      │
│  .... about 71% of women between the ages         │
│  of 20 and 24 were already married.  By 1978,     │
│  the proportion had dropped to 52%.  The          │
│  longer childbearing is postponed, the fewer      │
│  fertile years a woman has remaining.             │
│                                                   │
│  The decline is not new.  The birth rate has      │
│  been coming down for the past 200 years in       │
│  the US - with the exception of one period.       │
│  The real question, and the more perplexing       │
│  one, is what caused the exception - the          │
│  baby boom that lasted for more than a decade     │
│  after World War 11..                             │
│  NY Times 3.12.78.                                │
│                                                   │
│  For more Social Sciences - Key #                 │
└─────────────────────────────────────────────────┘
```

Figure 12

We do not know how our new public will take to Prestel or to Scitel in particular. We do know that the average rate of page consumption is about four per minute; this will cost a viewer about 3p or 6 cents in telephone time and Post Office service charges; additionally each information provider will levy his own per page charge, with the Post Office collecting and distributing the revenues. In some cases pages will be free - for instance with holiday tour information or cars for sale, revenue will be received from advertisers. Assuming that the total cost is 5p or 10 cents for 4 pages per minute, a viewer could participate in about five 16-page sessions five times a week for about £50 or $100 a year - the price of a good quality daily newspaper. The modified television receivers are expected to cost about £50 or $100 more than standard receivers when they are in mass production.

Future Possibilities

It is interesting to speculate about the probability of take-off for an innovation of this kind. In Figure 13 two scenarios are compared; technological success is only one of several necessary ingredients. The optimistic predictions made for U.S. interactive Cable TV in 1969 and 1970 were not fulfilled for the reasons shown. The conditions for Viewdata/Prestel in the U.K. appear to be more favorable.

```
┌─────────────────────────────────────────────────────┐
│                                                       │
│              INNOVATION CONDITIONS                    │
│                                                       │
│    U.S.    Interactive Cable TV                       │
│            Content dilemmas                           │
│            No standardisation                         │
│            Cost of entry                              │
│            FCC regulations                            │
│            RESULT - NO TAKEOFF                         │
│                                                       │
│                                                       │
│    U.K.    Viewdata                                   │
│            Monopolies/quick standardisation           │
│            Current technology                         │
│            Cheap revenue collection                   │
│            All-purpose terminal                       │
│            Simple indexing/cheap computer             │
│            Content right?                             │
│            RESULT - TAKEOFF?                           │
│                                                       │
└─────────────────────────────────────────────────────┘
```

Figure 13

Developments for similar services are in hand in a number of countries. In the U.S., Insac of New York, on behalf of the U.K. National Enterprise board, is considering implementation using the Post Office software with variations. They envisage terminals costing about $500 and computer connect time at around $1 per hour. Insac believe that the combination of the GEC (not to be confused with GE) highly economic minicomputer together with the Post Office's comprehensive software, with more developments on the way, is a competitive force to be reckoned with.

The regulatory position in the U.S. is very different from the U.K.'s monopoly system, and the situation is not yet clear. It seems likely that AT&T would be permitted to do no more than act as a carrier for Viewdata signals, but there seems to be no reason why suitable consortia should not be formed using available communication network facilities. On the other hand, I would expect the competitive climate prevailing in the U.S. to erode the British lead quite rapidly. The British scene is almost dictatorial by comparison; this has enabled the state monopoly to make the investment and set standards at a rate which would be impossible in the U.S. But Texas Instruments is experimenting with a Ceefax-like system in Salt Lake City, and GTE has been working on home information systems for the last two years.

The course of future events may be substantially modified by the integration of home computers into Viewdata-like

systems as was mentioned at the beginning of this paper. Included in this integration would be cheap printers for page
reproduction attached to the home terminal - and this immediately raises the possibility of arranging an "electronic mail"
system. In fact Prestel has this facility built into it - on-
screen messages of a rather elementary kind can be exchanged
between viewers now, experimentally. If full keyboards were
added complete messages could be exchanged.

British television equipment manufacturers are actively
considering the connection of home computers to receivers. A
receiver, embodying all the additional electronics, including
a microprocessor, has been built to accept computer programs
broadcast via Teletext. A program, stored for as long as it
may be needed in the receiver's existing page storage, is ex-
ecuted by the microprocessor, and the TV screen is used for
display. Such an arrangement would enable a viewer to run-in
whatever program he needs; one idea which has been demonstrated is a "teach yourself first-aid" interactive program.
At the same time the full power of access to very large information files via Prestel would be available to the viewer
using the same all-purpose terminal. This is another step
towards the "Consumersole" which I have described elsewhere.*
Many activities, such as the purchase of goods and services
and payment for them without leaving home would be possible.
This brings with it a host of social problems. Amongst the
less publicised of these are an acceptance of the equivocal
but persuasive authority of the tube, and so the information
conveyed upon it, and a further deterioration in people's
capacity for good conversation and democratic controversy.
Charles Ferris, chairman of the Federal Communication Commission, addressing a White House Press briefing in November
said, "Lots of creative things could be done with Viewdata
and Teletext systems, but the consumer hasn't thought of doing
them yet." I hope that the social consequences will receive
at least as much attention as the technology.

---

* Cawkell, The Paperless Revolution. Wireless World.

Bibliography

Cawkell, A. E. Developments in interactive on-line systems and teletext information services in the home. On-Line Review. 1(1): 31-38; 1977.

Cawkell, A. E. The paperless revolution. Wireless World. 84(1511/1512): 38-42, 69-74; 1978.

Fedida. S. Viewdata - a Post Office interactive information medium for the general public. Electronics & Power. 23(6): 467-473; 1977.

Fedida, S. The Viewdata computer. Wireless World. 84(1508): 44-48; 1978.

Garfield, E. Viewdata and Scitel bring inter-active information systems into the home. Current Contents. 1977(40): 5-10; 1977.

Hedger, H. Telesoftware - home computing via Teletext. Wireless World. 84(1515): 61-64; 1978.

Weizenbaum, J. Once more a computer revolution. Bull. Atomic Sci. 34(7): 12-19; 1978.

Viewdata - a review and bibliography. On-Line Review. 2(3): 217-224; 1978.

The Prestel user's guide and directory - a complete guide to information providers and services (Price £1,50). Eastern Counties Newspapers, Rouen Road, Norwich, England.

Part 2

Operational Trials of Electronic Information Exchange System (EIES)

4. An Overview of the Nature, Purpose and Initial Findings

Abstract

In the late fall and early spring of 1977-78, four "small research communities" began using EIES, a computer-mediated communications system. The purpose of the operational trials was to see if the use of a system of this type to tie together geographically-dispersed scientists working in the same research specialty area would increase the communication and productivity within the groups. All four groups were charged with assessing the impact of the system upon their members. In addition, the Division of Mathematical and Computer Science, NSF, funded an overall study which provides comparable data across groups. This paper summarizes the data collected in the "pre-use" and "three-month follow up" stages of the project. The distribution of usage, reasons for low use, and subjective evaluations of the system by its users are presented.

Introduction

The Electronic Information Exchange System (EIES) was designed to enhance communication within geographically dispersed "small research communities", "conceived as groups of 10 to 50 individuals sharing an interest in a scientific or technological problem area." (1). EIES provides a message

This project is partially supported by the Division of Mathematical and Computer Science, National Science Foundation (NSF-MCS-7727813). The EIES Operational Trials are supported by the Division of Information Science and Technology, NSF. The opinions and findings reported here are solely those of the author.

system which enables members to send private communications to individuals or groups on the system, "conferences" which build up a permanent transcript on a topic of discussion, and note-books where scientists may use text processing features to work on jointly authored reports. It also provides the capa-bility for the publication and dissemination of "electronic journals." (See Hiltz and Turoff (2), for much more detail on the design and capabilities of this and similar systems.)

The Division of Science Information (now the Division of Information Science and Technology) of the National Science Foundation issued a program announcement in 1976 inviting pro-posals for "operational trials" of the system. Four groups were chosen to participate, beginning in late 1977; three more groups were chosen in 1978. This is an initial report on the use of the system and reactions to it, based on the first four scientific communities supported under the operational trials program, plus two other groups of scientists who used the sys-tem during the same period.

The official objectives of the operational trials program were quite modest:

> to test the hypothesis that electronic information exchange (EIE) can enhance the effectiveness of individuals belonging to a small research com-munity,

> to accumulate practical experience with EIE by the members of such a community, and

> to gain deeper insight into the relationship between communication processes and the progress of science and technology. (1).

On the other hand, the designers and the principal in-vestigators of the scientific groups hoped for rather dra-matic and perhaps even revolutionary impacts stemming from increased communication due to use of the system, in the sense of Kuhn's concept of "paradigm shifts" constituting scientific revolutions. (3). This was particularly true be-cause all four of the initial groups represented young, inter-disciplinary fields which could not be said to have previously developed an agreed-upon set of theories and methods to guide their research.

The Division of Mathematical and Computer Research funded a study by the author of this paper to conduct an

across-groups assessment of the impact of the use of EIES,
with the following objectives:

> feedback to the designers on user reaction
> to specific features of the system,
>
> isolation of the factors accounting for low
> vs. high levels of use, and
>
> identification of the individual and group
> impacts of the system on those who do make
> substantial use of it.

Data were collected by a pre-use questionnaire and 3-
months follow up mailed to each member of the target groups
and automatic monitoring of usage statistics, plus personal
and on-line interviews to supplement the structured interviews. It is the initial results of these data sources which
form the basis for this paper. Final follow-up interviews,
analysis of the content of conferences conducted on the system, and other qualitative materials have not yet been collected or analyzed. Thus the data reported here are only a
small portion of the data which eventually will be analyzed
and reported on for this project. However, it is time for the
scientific community at large to be informed of what has been
happening thus far in the operational trials.

The User Groups

Group 30, Futures Research Methodology, is composed of
persons who have conducted planning, forecasting and similar
studies, and are attempting to discuss and improve methodology
in this area.

As pointed out in the proposal submitted for this operational trial, "Since futures research methodologists come from
a wide variety of backgrounds and disciplines, the channels of
communication which would ordinarily be provided by a single
professional society do not exist." (4). It was hypothesized
that use of EIES would significantly enhance the rate of innovation and dissemination of fruitful new ideas in the field.
These conditions and hopes are similar to those stated in the
other three proposals, for groups 35, 40, and 45.

Group 35 is the Social Networks Community, which is composed of sociologists, anthropologists, political scientists,
and others who share an interest in the study of social networks, or the patterns and types of "ties" that connect

members of groups or communities of various types. As they
state in their self-description, their aim is to "enhance individual productivity and to facilitate the development of
group goals, standards and the like."

Group 40 is General Systems Research. As their principal
investigator states, "General System theorists constitute one
of the few research communities that is deliberately trying to
integrate a wide variety of scientific disciplines. ... The
group plans to use the test facility not only to conduct research but also to educate each other in the various disciplines and approaches involved. As 'common tasks' the participants will compile a glossary of terms and a disciplinary
matrix for the field." (5).

Group 45 consists of people who share an interest in the
development of assistive and adaptive devices for the disabled, and includes disabled persons, research engineers, and
consumer-oriented organizations.

There are also two smaller groups included in the study
which have used the system on an unfunded basis. Group 50
used the system for about three months in the spring of 1978.
It consisted of about a dozen computer scientists and information analysts interested in the use of systems like EIES
for information analysis tasks. An interesting aspect of
Group 50 is that they "moved" to EIES from the PLANET conferencing system.

Group 80 consists of medical doctors in the area of
hepatitis research who are attempting to build a data base
about hepatitis which could be used by other doctors. They
received a very simplified version of the system and were
instructed to use only part of its range of capabilities, in
order to cut learning time to a minimum.

Finally, there are three new operational trials which
began in the summer or fall of 1978. These are Legitech, a
network of state legislative science advisors; a group studying mental workload and planning to create and publish an
electronic journal on the EIES system; and JEDEC, a set of
standards-setting groups sponsored by the Electronic Industries Association. No quantitative follow-up data are yet
available for the newest groups. However, short reports on
their experiences thus far will be included in the papers
which follow this one.

Amount of Use of the System and its Determinants

Table 1 shows the distribution of total amount of use of the system by the six groups, as of November 1978.

About a third of the scientists invited to use the EIES system either never signed on at all or used it so little that they never got through the learning period to the point where they were exchanging communications with other members of their group. The majority used it a disappointingly modest amount, between five and fifty hours, and assumed a mostly passive or "read only" role in the conferences.

Why this disappointing level of adaptation to and participation in the system? Table 2 shows the factors reported as being very important in limiting the use of the system.

Table 1. Total Hours of EIES Usage, by Group, as of November 1978

Group	<5	5-49	50-99	100+	Total
30	32%	39	19	10	100%,N=31
35	24%	60	7	9	100%,N=42
40	37%	48	8	7	100%,N=60
45	47%	47	4	2	100%,N=47
50	45%	45	10	0	100%,N=11
80	0	86%	7	7	100%,N=15
All,N	69	106	18	13	206
	34%	51%	9%	6%	100%

Source: EIES on-line usage statistics. Only persons with ID's in the ranges indicating primary membership were included. The persons included above had joined the system from one to twelve months previously, with most memberships beginning in January 1978. Note that the data slightly underestimate the number of "false starts;" a small but not recorded proportion of the ID numbers in each group had been replaced by new members by November.

Table 2. Percent Checking Reasons as Very Important in
Limiting Use of EIES, Spring 1978

Other professional activities must take higher priority	54
Limited night or evening hours	29
Inconvenient access to a terminal	19
Trouble with Telenet	18
Had some bad experiences	15
Cost of telephone or Telenet	13
Trouble with telephone	8
There is no one on this system with whom I wish to communicate a great deal	8
The conference comments or messages I have received do not seem worth reading	7
I do not know how to type	7
The system is too complicated	7
Inadequate leadership of the group	6
I am not very interested in the subjects being discussed	5
Red notebook documentation looked like too much to read	5
I do not like using a computer system like this	3

Source: Follow up questionnaires sent to groups 30, 35, 40,
45, 50 and 80 in spring 1978. Total N responding is 117.

In analyzing this table, the theoretical framework that is used has been provided by Robert Merton and other scholars in the area of the "sociology of science." This means that use of EIES is seen as an element that must be fitted into the social organization of science as a whole, with its reward structure of particular importance. (See Merton (6), for a good overview of this theoretical approach).

A companion theoretical approach is provided by George Homan's "exchange theory," which points out that people will only engage in behavior for which the "rewards" outweigh the "costs." (7). What we will do, then, is look at some of the costs in time, frustration, or money that are perceived by EIES members, and how this compared to the rewards of using their time on the system rather than in other professional activities.

Barriers

There are certain conditions which make it highly likely that a person invited to use EIES will use it little if at all.

First there is access to a computer terminal. Ideally, everyone would have his or her own terminal, at home as well as at work. In reality, some members of the system had access only to a terminal someplace else on campus, shared with many other users. This provides in effect very little access at all.

Many users could not get through frequently inadequate and unreliable TELENET service to access the system, or had interference and difficulties with their telephone lines. The telephone difficulties were bureaucratic in many cases. For instance, a Swedish user reports that his university requires him to obtain a special outside line for which he must wait about a half hour each time. An American reports that his university operators regularly cut off his WATTS line after half an hour, before he wishes to end the session. Then there is the possibility of experiencing a crash or slow response if the system is reached.

As a new medium of communication, there must be a considerable investment of time in learning to use the system before the medium becomes usable at all. In our follow up sample of users who had spent at least five hours on line, seventy five users reported a mean of 4.9 hours (median, 3.5, mode is two) to "learn to use EIES reasonably well." This is higher than the learning time reported during the pilot period (8), which is reasonable because the complexity of the system had grown and many more users had never had any experience with interactive computer systems of any type.

Given these formidable barriers to becoming comfortable with the system, it takes considerable motivation for a user to get through the learning period to the point where EIES seems to be "worth the trouble."

However, by far the most important factor is lack of motivation on the part of the scientists themselves. It is not that they do not want to talk about other people's research or engage in theoretical or methodological debates. It is that this simply cannot compete in terms of the priorities, with the press of immediate professional demands of their jobs. By far the most important reason given for limiting use of the EIES system is that "Other professional activities must take higher priority."

Table 3, below, shows the cross tabulation of the importance of other activities having priority vs. total use of the system at the time the question was asked, the end of May 1978.

Conversations on-line and qualitative comments written on the questionnaire also confirm that most members of the research communities on EIES feel that their activities there are not "real work" and that they can receive little in the way of professional advancement as a result of their contributions to group discussions or activities. Thus it would

Table 3. Relative Priority of Other Work vs. Actual Use of EIES

	Less than 5 hrs.	5 to 49	50 and over	N
Very Imp.	41	49	10	61
Somewhat	22	50	28	36
Not Imp.	11	56	33	18

Gamma =.45

Chi Square =11.41, p=.02

seem that in the future, groups which use systems of this sort should have a task to accomplish which is defined by its members as a high priority professional activity, whose successful completion is important to their on-the-job advancement.

Group 80 provides some support for this interpretation, though of course there are many other possible interpretations of the reasons for differences among groups. It will be remembered that this group is using the system to actually carry out an important research project. Only one of the group 80 members checked the priority of other professional activities as being very important in limiting use of EIES, and all of group 80 as of November shows at least moderate amounts of use.

Another strong predictor, along the same lines, is whether or not "There is no one on this system with whom I wish to communicate." The gamma on correlation with actual level of use for this factor is .52, and it is significant at the .04 level.

Here are some of the qualitative comments contributed by respondents which emphasize the importance and frustration of this motivational factor.

(From a group 35 member who checked that EIES is "frustrating")- "Not so much EIES as the pressure of other activities that keep EIES out."

"EIES is still a spare time activity and not part of my regular professional activities." (group 30)

"I have no negative comments about the system. My problems in use derive almost entirely from my own difficulty in typing and the press of more urgent matters." (group 30)

"Just not enough time to devote to conferences, not EIES fault." (group 30)

"I do not have enough time to work the EIES.. I also have little time to do 'creative thinking' as a staff therapist with most of the time set aside in case sessions." (group 45)

"High overhead to learn the system... requires TIME to learn how to use the system effectively and efficiently; this was a big barrier initially, as other tasks/functions took priority." (group 50)

Finally, we see that the reason which comes out as the second "most important" in limiting use of the system is limited night hours. If a scientist considers communication with his or her peers something which is not part of "work" but which must be fitted into "spare time," then it becomes important to be able to use the system from home during the evening, rather than having to fit it into the work day at the office. As one frustrated group 35 member put it who had only a shared terminal at the office, "I wished I could really have used the system, but it simply cannot compete with other demands on my university office time."

In the near future, EIES should be up around the clock, and this should improve usage opportunities for those scientists who see communication with their peers in other locations as an activity of secondary rather than primary importance.

Perception of Competitive Factors as Related to Subsequent Use of EIES

Another motivational factor related to use of EIES is a scientist's perception of the nature of competition in the specialty area. Let us put this into context with a quote from Merton (9):

> The values and reward-system of science, with their pathogenic emphasis upon originality, help account for certain deviant behaviors of scientists: secretiveness during the early stages of inquiry lest they be forestalled, violent conflicts over priority, a flow of premature publications designed to establish later claims to having been first. These, I suggest, are normal responses to a badly integrated institution of science, such that we can better understand the report of a sample of American 'starred men of science' that, next to 'personal curiosity', 'rivalry' is most often the spur to their work.

This is not to suggest that failure to use EIES is "deviant"; however, there is definitely a relationship between the perceived nature of competition and rivalry and whether a person is likely to wish to use EIES to share ideas and preliminary research results, as shown in Table 4.

Table 4. Perceived Reasons for Competition and Subsequent
Use of EIES

QUESTION: What are the reasons for this competition (check
all that apply)...

Checked	<5	5-49	50+	TOTAL
Scarcity of or Competition for Funds				
yes	33	52	15	100%
no	11	57	31	100%
X2=5.07,p=.06				
Some Persons Act Unethically				
yes	71	29	0	100%
no	16	57	26	100%
X2=11.3,p=.003				
Strongly Opposing Views				
yes	21	37	50	100%
no	22	61	16	100%
X2=5.37,p=.07				

Source: Asked at pre-use, most generally January 1978.
Dependent variable is total hours of use as of December 1,
1978.

The last cross-tabulation in Table 4 is most interesting.
If participants perceive a great deal of controversy within
their specialty, they are more likely to participate heavily.
However, if unethical behavior is perceived, then the probability of participating a great deal on EIES appears to be
zero; one must trust one's colleagues in order to share ideas
with them.

Subjective Satisfaction

In this section, we will report just a small portion of
the available data on reactions to the system by those who did
spend a minimum of five hours using it. This is the subgroup
which received a "long follow up" questionnaire probing their
detailed responses to aspects of the system.

Table 5.   Reactions to Specific Features of the EIES System

QUESTION: How valuable or useful do you currently find each
of the following features or capabilities of the Electronic
Information Exchange System for your own communications activity?   (If you have not actually used a feature, please check
"cannot say").

Feature	Extremely Valuable	Fairly Useful	Slightly Useful	Useless	Cannot Say
Private Messages	64%	31	3	1	1
Group Messages	23%	41	21	9	7
Group Con- ferences	22%	42	19	7	10
Private Con- ferences	30%	24	7	7	32
Public Con- ferences	14%	28	27	11	20
Notebooks	13%	23	6	7	51
Directory	24%	31	21	6	17
Retrieval Capability	22%	18	13	6	41
Text Editing	28%	25	15	6	25
Anonymity or Pen Names	4%	10	9	29	49
Explanations File	9%	29	19	11	33
+sen	12%	9	8	1	70
String Variables	17%	6	6	3	68

Source: Three month follow up surveys to groups 30, 35, 40,
45, and 50.   Total N responding is 72.

Table 5 shows an unanticipated set of usage and reaction patterns. Whereas the primary activity on the system was envisaged to be group conferences, this in fact is seen by most members of the system as less valuable than the one-to-one or one-to-a-few communications which take place in the private messages. One also sees an indication that the large group conferences are seen as somewhat less satisfactory than smaller "private" conferences set up by individuals to discuss more narrowly defined topics of interest to a subgroup of people on the system.

We also see that the majority of members of the system, as of May or June, had never used some of the more advanced features of the system, such as the "string variables" which allowed a person to custom program interaction sequences. Most of the members of the system had never tried using a notebook and almost half had never even tried using a pen name or anonymous sending of a message.

Those who did use the system tended to give it fairly positive, but not "perfect" ratings, in terms of their subjective reactions to the experience of using the system. Subjective reactions are, needless to say, highly correlated with total amount of use of the system.

For example, one of the subjective rating scales was asked and answered as follows, at the three month follow up:

Overall, the EIES communication system is

```
:   1   :   2   :   3   :   4   :   5   :   6   :   7   :
Extremely                 Neutral                 Extremely
Good                                              Bad
```

When cross tabulated by total hours of use in June, the correlation (gamma) was .54, significant at the .001 level.

Mean values for the one-to-seven rating scales are shown in Table 6, along with the distribution of responses. The most favorably rated aspects of the EIES system are that it is stimulating, fun, friendly, and easy to use. The most mediocre ratings are in terms of its being frustrating, time wasting, unproductive, and intrusive. Based on the data summarized above, it is not so much the system as the lack of a concrete shared research project or other "real work" to accomplish which seems to account for the system's being seen as unproductive by many of the members of the initial operational trials groups.

Table 6. Overall Reactions to the EIES Mode of Communication

: 1 :	2 :	3 :	4 :	5 :	6 :	7 :	Mean
Extremely Good						Extremely Bad	
12%	42%	30%	3%	9%	5%	0	2.7

: 1 :	2 :	3 :	4 :	5 :	6 :	7 :	
Stimu-lating						Boring	
19%	39%	29%	6%	6%	1%	0	2.4

: 1 :	2 :	3 :	4 :	5 :	6 :	7 :	
Produc-tive						Unproduc-tive	
7%	22%	39%	15%	11%	3%	3%	3.2

: 1 :	2 :	3 :	4 :	5 :	6 :	7 :	
Great Fun						Unpleas-and Work	
15%	38%	22%	13%	11%	1%	0	2.7

: 1 :	2 :	3 :	4 :	5 :	6 :	7 :	
Time-Saving						Time-Wasting	
9%	20%	19%	24%	21%	3%	6%	3.6

: 1 :	2 :	3 :	4 :	5 :	6 :	7 :	
Not Frus-trating						Frustrat-ing	
7%	14%	26%	21%	19%	7%	6%	3.7

: 1 :	2 :	3 :	4 :	5 :	6 :	7 :	
Friendly						Impersonal	
11%	35%	33%	14%	4%	3%	0	2.8

: 1 :	2 :	3 :	4 :	5 :	6 :	7 :	
Easy						Difficult	
15%	32%	26%	15%	10%	1%	0	2.8

: 1 :	2 :	3 :	4 :	5 :	6 :	7 :	
Not demand-ing or intrusive						Very de-manding or intrusive	
9%	24%	23%	20%	23%	1%	1%	3.3

Source: Long follow ups, N=72

Conclusion

The EIES system appears to be a viable means of facilitating communication within a scientific community. However, it does require convenient access to a computer terminal and high motivation on the part of its users to communicate with one another. In order for users to invest the time to learn to use the system and to participate actively, the activity which they are carrying out on the system must be seen by them as important and rewarding.

Acknowledgments

I would like to thank Mary Ann Solimine for her diligent work in obtaining the SPSS runs on which this paper is based; and the student assistants who helped in the coding and data preparation for this project: Gerry Lavner, Chris Naegle, Nancy Wymer, Joanne Garofalo, Stacy Simon, Keith Anderson, Elizabeth Hunt, Dorothy Preston, Nancy Rabke, and Ned O'Donnell.

References

1. National Science Foundation, Division of Science Information. Program Announcement. Operational trials of electronic information exchange for small research communities. NSF76-45; 1976.

2. Hiltz, Starr Roxanne; Turoff, Murray. The network nation: human communication via computer. Reading, Mass.: Addison Wesley Advanced Book Program; 1978.

3. Kuhn, Thomas S. The structure of scientific revolutions. (Second edition, enlarged). Chicago: The University of Chicago Press; 1970.

4. Martino, Joseph. A computer conference on futures research. Proposal submitted to National Science Foundation; 1977.

5. Umpleby, Stuart. General systems theory: an example of the integration of scientific disciplines. Proposal submitted to National Science Foundation; 1977.

6. Merton, Robert. The sociology of science: theoretical and empirical investigations. Edited and with an introduction by Norman W. Storer. Chicago and London: University of Chicago Press; 1973.

7. Homans, George. Social behavior as exchange. American Journal of Sociology. 62; 1958.

8. Turoff, Murray; Hiltz, Starr Roxanne. Development and field testing of an electronic information exchange system: Final report on the EIES Development Project. New Jersey Institute of Technology, Computerized Conferencing and Communications Center. Research Report Number Nine; 1977.

9. Merton, Robert. The ambivalence of scientists. Sociological Ambivalence and Other Essays. New York: The Free Press; 1976.

5. Computer Conference on General Systems Theory: One Year's Experience

Introduction

The computer conference on general systems theory has developed a glossary for the field, begun a history and a genealogy of theorists, reviewed topics of current interest and built ties of informal communication among people in the field. A paradigm debate started in the main conference has sparked controversy and curiosity in addition to confusion. Despite these substantive achievements, a computer conference can be a frustrating experience. Frequent users are disappointed by the slow replies to their messages and comments. Infrequent users are overwhelmed by the massive number of messages and comments waiting for them when they sign on. It is apparent that a computer conference is a unique new form of communication. Effective use requires regular use, but this is currently the exception rather than the rule.

Learning to Fly

Among the principal investigators of EIES groups there seems to be a feeling of disappointment in the results of the exercise so far. Nevertheless these people still believe in the potential of computer conferencing, so they tend to search around for explanations which will put an acceptable face on disappointing results. In this state of mind I came across the following paragraphs in the November, 1978, issue of SCIENTIFIC AMERICAN, p. 167.

Access to the Electronic Information Exchange System (EIES) is provided by the National Science Foundation's Division of Information Science and Technology. The opinions and conclusions are those of the author and do not necessarily represent those of the National Science Foundation.

Wilbur and Orville Wright invented the controllable
airplane. Until they first flew in public in 1908
it was believed a powered aircraft would be similar
in its behavior to an airship, a stable vehicle that
could be steered right and left by a rudder and up
and down by a horizontal rudder, elevator. One could
expect to mount such a craft and fly it without previously acquired skills, and that was what was invariably attempted until the Wrights showed how it
should be done.

In contrast, the Wrights, who were builders of
bicycles, conceived the airplane from the beginning
as being a vehicle that like the bicycle depended
on its operator not only for its direction but also
for its equilibrium. It therefore seemed perfectly
natural to them that before one could hope to successfully operate a powered aircraft one needed to
develop both the aircraft and the skills necessary
for operating it.

There seems to be an analogy here to the operation of a
computer conference. I for one began with the assumption that
a computer conference should pretty much take care of itself.
If a group of people with a common set of interests were given
access to EIES, I expected that they could conduct their normal professional communication with enhanced speed and effectiveness. Alas, this was not to be. A few months into the
conference, helpful user consultants began referring to previous studies of conferencing behavior which concluded that
"strong leadership" was necessary for the success of a computer conference. I strenuously resisted this suggestion.
Not only did it offend my democratic sentiments, it implied
more work! But the evidence seemed to support the need for
strong leadership. Hence I embarked upon a strategy of dele-
gation of authority. Surely several strong leaders were
better than one. Asking others to serve as moderators of
conferences I had set up was not very successful. But a few
people functioned very effectively as moderators of conferences they had initiated, for example Roger Conant, Bill
Halal, and Jay Forrester.

It appears that an active moderator is necessary to keep
the conference going but that as people get used to the system
and initiate their own projects, several leaders begin to
emerge. "Learning to fly" on EIES requires a fair amount of
practice. It takes people about 10 hours to learn to use EIES
well. In about 100 hours they become skillful communicators
and organizers within the EIES community. Perhaps about 1000
hours are required to learn, through EIES, a field not previously intensively studied.

Group Process

As shown in Figure 1, the usage of the system follows the academic year, with August being the least active month. It appears that the moderator has to "restart" the conference at the beginning of each semester or the level of usage drops off. This is another hard-to-delegate task for the moderator.

```
Figure 1

HOURS USED BY GROUP 40

          100         200         300         400         500

Febru ***************
March *************************
April ***********************
May   ***********************
June  *******************
July  ***********
Aug   ******
Sept  ***********
Oct   ********************
Nov   **************
Dec   **************
Jan   *******************
```

The distribution of personal activity has been highly skewed. Figure 2 shows the activity of individuals in conference 40, the most active discussion conference of group 40. The numbers down the left hand side indicate people. The moderator is number 001. It may be that the activity of the moderator determines the activity of the group in a manner that is fairly predictable.

It should also be noted that during the fourteen month period, November 1977 through December 1978, the medium rate users of group 40 entered less than one comment per month in conference 40. Although conference 40 was not the only conference the group was using, it was by far the most active conference. Only four people entered an average of one comment per month or more.

The activity in a conference over a period of months is more skewed than the activity during a single month. Figure 3 shows the activity in conference 40 during a typical month.

```
                            Figure 2

               INDIVIDUAL ACTIVITY IN CONFERENCE 40
                  November 1977 to December 1978

                   NUMBER OF COMMENTS PER PERSON

           10         20         30         40         50

001 **************************************************
002 ********************
003 ***************
004 **************
005 ************
006 ************
007 ************
008 ***********
009 ***********
010 *********
011 *********
012 *********
013 *******
014 ******
015 ***
016 ***
017 ***
018 ***
019 ***
020 ***
021 ***
022 **
023 **
024 **
025 **
026 *
027 *
028 *
029 *
030 *
031 *
032 *
033 *
034 *
```

The order of people and their numbers is the same as in Figure 2. The highly skewed distribution in Figure 2 is therefore more the result of regular versus intermittent activity rather than the result of regularly high or regularly low use. It is apparent that a computer conference is a unique new form of communication. Effective interaction via EIES requires regular use, but regular use is currently the exception rather than the rule.

Figure 3

INDIVIDUAL ACTIVITY IN CONFERENCE 40
During March 1977

NUMBER OF COMMENTS PER PERSON

	10	20	30	40	50
001 *****					
002					
003 *					
004					
005 ***					
006					
007					
008 ****					
009					
010					
011 **					
012 ***					
013 **					
014 *					
015					
016					
017					
018 **					
019					
020					
021					
022 **					
023					
024					
025					
026					
027					
028					
029					
030					
031					
032					
033					
034					

```
                            Figure 4

        ACTIVITY IN CONFERENCE 40 BY PERSON AND BY MONTH

     nov dec jan feb mar apr may jun jul aug sep oct nov dec
001 ********____*********************_____*********** 
002                     ****************
003               *************************____****
004 ****_____*******_____****
005       ****____****____***********
006                           ****_____****
007                              ****_____****
008     ****************
009         ****_____****_____****
010 ********_____****_____****
011         *********************_____****
012            ****____********_____****
013       ****____********____****_____****
014     ****************
015   ********_____****
016       ********_____****
017       ********
018           ********
019                   ****_____****
020                                               ****
021                                               ****
022             ****
023                 ****_____****
024                   ****
025                                               ****
026 ****
027       ****
028       ****
029       ****
030               ****
031                 ****
032                 ****
033                                               ****
034                                               ****
```

The fact that regularity of use is the principal cause
of the skewed distribution of participation is further illus-
trated by Figure 4. Four asterisks under a particular month
indicate merely that one or more comments were entered that
month. Dashes mean that due to subsequent comments, the
person is assumed to have read the intervening comments and
not dropped out. The moderator, 001, was active during the
entire period. Person 002 was very active from May to August
but then broke his arm and has not yet returned to the conference. Person 004 is active primarily between semesters.
Person 008 was active for a while but then encountered telephone difficulties at his university and dropped out. He
has since returned. Many people entered one or a few comments and then were not heard from again.

Topics Being Discussed

Below is a list of conferences used by group 40 and the
number of comments entered in each one by the end of January
1979.

Conference	Comments
40, general systems theory	273
41, glossary	282
42, genealogy and history	20
43, science policy	10
44, communication networks	68
400, disciplinary matrix	21
401, talk about talk	37
402, instructional tv	36
403, requisite variety	32
404, schemas and category theory	17
410, information	39
415, evolution	35
416, SGSR task force	25
418, system formation	77
426, conference panel	26
428, hierarchy theory	15
437, educational programs	257
438, course outlines	10
439, annotated bibliography	4
448, rural telenet users	5
449, test questions	62

European Participation

Early in the project I tried to get a number of people in Europe to join us. Unlike the U.S. and Canadian users they would have to pay their own communications charges. This effort has been more difficult than expected. European users face a number of difficulties ranging from unreliable hardware to institutional impediments. However by far the most serious difficulty is the price of communications services. In Europe government monopolies control the Post, Telephone and Telegraph (PTT). Needless to say these companies do not want their telephone systems to undermine their telegraph business. Hence they charge by the character, thereby eliminating any incentive to get more messages through the same equipment. The importance of these tariffs in stifling innovation is beginning to be recognized. The issue may come up at the UN conference on Science and Technology for Development in Vienna in August 1979.

In Britain, the use of EIES may be illegal. Apparently it is legal to use a computer network to access a data base but not to send messages. It was argued that EIES users were accessing a data base of verbal material. However, the British Post Office decided EIES was more a message system than a data base.

Writing a Handbook

During the first year of operation on EIES the group primarily engaged in open-ended discussion on topics of general interest. Several tasks were undertaken such as developing a glossary, history, and "disciplinary matrix" for the field. However, it was never clear what would be done with this material. During our second year we plan to begin work on a handbook of general systems theory. I am hoping that many of the members of the group will want to contribute to this undertaking. It will be interesting to see how the shift from open-ended discussion to concentration on a task affects the group's use of EIES.

Acknowledgments

The members of the EIES group on general systems theory are: Stewart Bainbridge, Jay Forrester, Barry Clemson, Roger Conant, Tom Wickenden, Joseph Goguen, Bill Halal, George Klir, Roger Cavallo, Klaus Krippendorff, Bill Gray, Irwin Marin, Terry Oliva, Larry Richards, Tom Simon, John Sutherland, George Tracz, Len Troncale, Francisco Varela, Crayton Walker, Zavis Zeman, Martin Zwick, Michael Arbib, Barry Hughes, Manfred Kochen, Bill Reckmeyer, Rolf Wigand, Steven Rogers, Anders Sandberg, Lucien Gerardin, Joe Martino, Murray Turoff.

References

Ashby, W. Ross. An introduction to cybernetics. London: Chapman and Hall; 1956.

Hiltz, Starr Roxanne; Turoff, Murray. The network nation: human communication via computer. Reading, Mass.: Addison-Wesley Publishing Company; 1978.

Kochen, Manfred, ed. Information for action. New York: Academic Press; 1975.

Umpleby, Stuart A.; Umpleby, Karen Thomas. Applying systems theory to the conduct of systems research. Paper prepared for the International Conference on Cybernetics and Systems, Amsterdam; 1978 August.

John Bregenzer, Joseph P. Martino

6. Futures Research Group Experience with Computerized Conferencing

Abstract

Futures research is an emerging discipline with rudimentary communications channels and a wide variety of backgrounds among participants. This leads to disparate theoretical and methodological stances, and sometimes even to considerable variation in the meaning of core concepts. The authors felt that computerized conferencing might coalesce the field, establish paradigms, make futures research more like established sciences. The first year of the conference has shown both enticing success in this direction and frustrating disappointments.

Introduction

When we applied for funding of our conference, we felt we had a good chance of success because of the nature of our group. Futures research could be thought of as a scientific and academic discipline in the process of birth. Members of this community are located at many universities, some not-for-profit research organizations, and some industrial and governmental planning and analysis activities. It is an active and productive. group. However, although there are four or five university-based research centers, these usually consist of less than half a dozen people and there are no major research centers of the type one associates with, say, high energy physics. Further, there are no academic departments of futures research, let alone strong departments to advance the discipline. All members of this discipline have been trained in and often achieved renown in some other discipline.

This report was funded by the National Science Foundation (NSF-DSI77-17920). The findings and opinions expressed here are those of the authors and do not necessarily reflect the views of the National Science Foundation or the members of the Futures Research Conference.

As might be suspected, communications channels in futures research are not well developed. There are increasing numbers of futurological congresses, but they are sporadic and no substitute for the regular meetings of a professional society. There are two core journals in the field, "Futures," published in Great Britain and "Technological Forecasting and Social Change," published in the United States. However, these journals seem to be a communications channel from futures researchers to non-futures researchers rather than a channel among members of this community. The Science Citation Index shows that most references to articles in these journals are made outside the field. Only newcomers cite established researchers in the two core journals. Established researchers cite work outside of the field and, of course, their own prior work. At present, letters and long-distance phone calls would seem to be the major communication channels in this scientific specialty. Therefore this particular group seemed ideally suited to benefit from the enhanced opportunity for communication offered by computerized conferencing. We felt that computerized conferencing might bring about coalescence and advancement of this productive, but fragmented, diverse, and farflung community.

Futures researchers have become members of our conference in a way that we would describe as "self selection." In the first stage, the moderator invited people he felt would be good, active participants. Nonactive initial members were replaced with new members who either applied to the moderator or were otherwise selected by him. Sometimes the replacements had to be replaced in the same manner.

Types of communications on the system include private messages, the main group conference and private conferences on specialized topics. Private messages are the most frequent type of communication. As an experiment, we asked members of the conference to send copies of their private messages to the conference evaluator. Since most members did not do this we can do very little analysis of that use of the system.

Results

Shortly after we began, two private conferences were established on topics of concern to the futures research community: cross-impact analysis and structural modeling. These conferences resulted in considerable clarification of concepts related to these subjects. The conference on cross impact was particularly helpful in clarifying concepts related to that topic. It also provided a means for useful debate on the Baysian vs. non-Baysian nature of probability adjustments, and

the nature of time-dependent probabilities. The structural
models conference provided a vehicle for discussing the pro-
liferation of similar but not identical models in wide use.
Interest in both conferences waned, with neither reaching much
of a conclusion. One was discontinued, the other continues at
a low level of activity.

A private conference on decentralization of society was
started. This attracted considerable interest and participation from both within and outside our group. It was a
lively and enjoyable conference, providing a vehicle for much
exchange of ideas, but not reaching any conclusions.

Private conferences for the purpose of writing book reviews have been established twice. One has been in session
for over six months and has not yet concluded its activities.
The other has been in session for a lesser time, but has not
completed its work either.

The main conference was used for many things. For one
example, it was used to discuss issues related to the World
Futures Studies Conference (WFSF), prior to its being held in
Cairo, Egypt, in September 1978. Two members of our conference
attended this meeting. One transmitted the results of our
discussion to the WFSF, and placed a report on the meeting to
our conference.

In another example, in response to a request from one
member other members contributed ideas toward the development
of a history of the field of futures research, identifying
specific important innovations and the originators of these
innovations.

Other less weighty but innovative and important uses of
EIES included the following: transmission of manuscript reviews and book reviews to journal editors who were conference
members; requests for data and reprints; announcements of new
journals and newsletters; and travels of foreign visitors in
the U.S. to conference members, facilitated by rapid "turn-
around" of EIES communications, that is, avoidance of the
difficulties of coordinating phone calls between the eastern
and western time zones.

Disappointments

Perhaps we should be happy and rest content with the
substantial accomplishments of our conference. Being more
activist oriented, we would like to list four of our disappointments in the hope of solving the problems that seem to

be preventing computerized conferencing from being a truly
effective research tool.

1.   A surprisingly large number of people who requested
in writing to be in the conference have not participated at
all.

2.   Of those who do participate, a surprisingly large
number participate very little.

3.   There is very little discussion in depth. It is
seldom that conference comments provoke more than one or two
responses. We had expected chain reactions.

4.   Two cooperative book reviews have been started.
Here, because of the clearly defined goal, we expected to see
some real products emerge from the conference. But neither
has yet reached a conclusion, and one has been trying for six
months.

Toward Better Computerized Conferencing

Our experience with computerized conferencing is a mixed
bag of enticing success and frustrating disappointments.
Among the successes are conferences that clarified central
concepts and methodologies in the field, a lively conference
on the decentralization of society, and many communications
between individuals that were facilitated by the asynchronous
mode of this system. Computerized conferencing is much faster
than the mail and much less harassing and intruding than the
telephone.

Our disappointments could be summarized by saying that
getting active, committed participation in a conference like
ours is like pulling teeth. We do not blame the members.
They are proven active, enthusiastic futures researchers. We
do not here blame the technology, such as the real frustrations of computer breakdowns, and difficulties of learning to
use the system, including for some learning to type. We blame
the structure of the conference. Perhaps properly, it began
in an informal manner without clearly defined goals or an
agenda. Therefore members have been communicating as one
would at a cocktail party. Such communication is good, probably necessary to any group. But the focused, goal-directed
type of communication is sorely missed by some of us, and also
necessary to any group. That such communication can be accomplished is demonstrated by the quarterly reports to the National Science Foundation that we have written on the system.

Though we are at the same campus, we find it more convenient
to communicate over the computer than to try to coordinate
phone call or meeting times. This present paper was written
by both of us communicating over the system. These small
products demonstrate that discussion in depth and meaningful
results can be accomplished expeditiously through computerized
conferencing.

After One Year: Cautious Optimism

Overall, we are encouraged by recent levels and quality
of activity. One of the significant factors affecting participation, we believe, is the process of selection which has
been taking place during the existence of the conference. By
dropping inactive members from the conference and replacing
them with others, and repeating this as necessary, we have
built up a group of people who are interested in participating, and are highly motivated to get some results out of the
conference. We believe that this factor bodes well for participation during the remainder of the conference.

The level of activity is also beginning to show that EIES
is becoming integrated into the communication habits of the
participants. They are no longer treating it as an "add-on,"
but are turning to it as a primary means of communication with
those persons who are also participating.

During the first nine months or so, we were critical of
the shallowness of discussion. Few responses were made to
comments and exchanges were of short duration. This situation
appears to be changing. In October we had a productive month-
long discussion between four members. Beginning in December
we had a productive discussion of over a month involving eight
members. Whatever the reason for this change toward lengthy
discussion in-depth, involving a large number of members, the
authors of this report are pleased and the members of the
futures group should be pleased. The activity in the conference appears to be approximating what was envisioned when
it was proposed.

A decline of activity from October to December, followed
by a resurgence in January 1979 would seem to be related to
the pressures of the academic year, which have major peaks in
December and June. Our active members appear to enjoy and
profit from their participation in this experiment, but are
sometimes constrained from taking part by very real demands
on their time. Greater participation might require members
to participate in EIES instead of other activities. Recent

activity levels suggest that the rewards of EIES participation might increase to the point that members would choose EIES over other activities, e.g. collaborating on a project rather than doing it singlehandedly.

Computerized conferencing may be like many other innovations that start as toys and only later get down to business. Like a rookie football player, it shows a lot of potential.

Jane H. McCarroll

7. EIES for a Community Involved in R&D of Devices for the Disabled

Abstract

Innovative Systems Research, Inc. received a grant from the National Science Foundation in the fall of 1977 to assess the potential applications of the Electronic Information Exchange System by a multi-disciplinary community involved in the research and development of devices for the disabled. A trial network was established in January 1978 with the primary purpose of exposing members of the R&D community to the capabilities of the EIES and exploring its potential applications and impacts in the field. This paper summarizes the group's use of the system and some of the statistical data collected to date.

Introduction

It is generally recognized in the field of rehabilitation that a major need exists for greater accessibility of information, timely feedback, and more efficient means of communication among individuals and organizations involved in the application of technology for the handicapped. It was with this need in mind that Innovative Systems Research, Inc. (ISR) proposed to the National Science Foundation that an operational field trial be supported to assess the potential value of electronic information exchange to a device R&D community. The proposed project was approved for funding in the fall of 1977 and a trial user community established in January 1978.

The project reported on in this paper is funded by the Division of Science and Information Technology, National Science Foundation (NSF-DSI77-17924). The opinions and findings reported here are solely those of the author.

The project is conducted as an evolving "natural experiment," in that the community's use of the system is not subject to the artificial imposition of controls or limitations. The participants are encouraged to develop their own lines of communication and conference groups and to employ EIES's special capabilities in support of their current work, to share ideas and information with each other, and to explore different methods and techniques for making use of the system.

While the level of usage has not been as high as originally expected and frustrations have developed over the limited representation of professionals in the field and direct availability of pertinent information, some interesting activities have been initiated on the system and many potentially beneficial applications and improvements in the design and coordination of the network have been suggested.

Participants

The project community is a cross-section of approximately fifty organizations and individuals representing the device R&D spectrum of basic and applied research, marketing, production, distribution and utilization. The following characteristics of the device R&D community in general provided the rationale for the investigation of electronic information exchange as a viable means of improving and facilitating communications within this community:

- The interdisciplinary nature of activities and individuals

- The relatively small size of the overall community whose members are familiar with one another

- The exhibited willingness and desire of members to work with representatives of other disciplines and related interest groups

- The representation of a cross section of scientific and engineering disciplines, providing great potential for a generalization of results to other R&D communities

- The frequent direct involvement of users (i.e., disabled persons) in R&D activities

- Wide geographic dispersion

- Frequent difficulty attending meetings and conferences

- Common general goals

- A need to reduce R&D costs through avoidance of duplication of effort and greater dissemination of existing information

- Members' knowledge of communications theory and familiarity with information systems

Not secondary to the above factors is the social significance of the area. Although numerous innovative devices and systems to aid the disabled are continually being developed, many do not progress beyond the prototype stage. The reasons are many and include the general deficiency of information about the disabled population, the varied and unique needs of disabled individuals, and the diverse types of participants and test conditions necessary to the R&D process.

The trial network has experienced significant turnover in membership since its inception. A group of thirty participants was selected during the proposal and start-up stages, and each of those was asked to suggest potential additional members. Actual use of the system has varied greatly among the participants, with many never even signing onto the system. Inactive users have been replaced with new members, some of whom were "recruited" by the project staff upon members' request and many of whom had requested permission to participate.

Activities

The trial community members did not represent a cohesive group prior to their participation on EIES and many did not know each other and have not been involved in related efforts outside of their EIES group membership. In order for members to become acquainted with one another and to begin defining their objectives for the application of EIES, ISR conducted a two-day "in-person" conference in Philadelphia in early February 1978. The participants worked hard at defining what they believed would be meaningful applications of EIES for their research interests. The proposed activities fell into four categories as described below:

- A community newsletter containing information about project participants, community activities, upcoming events of interest to members, requests for information and suggestions for more effective use of EIES.

- A community notebook for dissemination of information about topics of general interest.

- Community interest conferences - i.e., group conferences on topics of interest to the group, including issues of law and public policy, device delivery systems, public information programs and strategies, interdisciplinary course development, and potential improvements in EIES performance and use.

- Special interest projects - i.e., private notebooks and conferences established by subgroups for specific purposes, including:

  - Exchanging data on a device undergoing clinical evaluation

  - Joint development of a standardized device evaluation form

  - R&D planning conferences

  - Special interest newsletters

  - Communication among developers and users of specific devices

  - Developing a linkage to an existing data base on devices

  - Preparing journal articles for publication

The limited scope of the project necessitated dependence upon group members to assume leadership positions in developing and coordinating many of the desired activities. ISR has initiated and moderated two group conferences - one a general forum for discussion of members' experiences with the system and a starting point for applications with well defined objectives. The second is an information exchange conference which members use to request specific information, to provide responses to such requests, and to announce upcoming events and other items of general interest to the group. This conference has been the center of greatest and most consistent activity within the group, apart from private message activity. Individual members have initiated and moderated the following activities:

- A temporary group conference to disseminate the agenda for the annual Conference on Systems and Devices

held in Houston, Texas in June 1978. This conference also provided a meeting place for the planning of an EIES exhibit at the conference by two
users from different locations.

- A private conference to discuss the needs of the
  rehabilitation engineering community for a data
  base on devices and to develop a preliminary
  categorization of devices.

- A private conference on the marketing and distribution
  of devices. The conference membership is interdisciplinary in makeup and an interesting and en-
  lightening discussion of the barriers to the com-
  mercialization of devices for the severely handicapped is still underway.

- A private conference for the exchange of information
  among the developer and clinical evaluators of a
  modular seating device. This conference was established to facilitate the reporting of clinical data
  and experiences to the developer of the device and to
  stimulate communications among the remotely located
  evaluation teams.

- A shared notebook which two members have used for
  preparing articles for a national newsletter on
  communication aids.

- A project evaluation conference between the project
  staff and NJIT personnel, to define the available
  statistics on system use and how they may be incorporated into the overall evaluation.

These activities actually represent a small percentage of
overall use of the system by the project community. It is
evident that most of the group's interactions are via the
message capability. ISR is currently obtaining permission
from members to collect and analyze message traffic, but the
content of all messages is private and confidential, making
it impossible to determine all of the uses to which the system
has been put. It is hoped that a better understanding of its
application by the group will be obtained through a questionnaire to be administered near the end of the grant period.

Preliminary Conclusions

The exploratory nature of the project has enabled ISR to
expose the network to a rather large and diverse audience in

the rehabilitation R&D field and, consequently, to define many
potential applications and network configurations to assist in
increasing their productivity through improved information exchange. The potential range of applications and effectiveness
of the medium, however, has hardly been touched upon by the
trial community. This can be attributed to many factors, the
most significant of which are probably the lack of direct access to existing data bases, the underrepresentation of professionals in the various relevant disciplines, the current
lack of interfaces and structures tailored to the specific
user community, and the short term funding period.

Participants to date have acknowledged a pressing, immediate need for better access to scientific and technical
information. Many of those who represent areas of rehabilitation technology (as opposed to manufacturers, clinicians and
consumer advocates) seem to feel, however, that the network in
its current stage of development requires more effort on their
part than benefits currently received from it. This situation
is necessitated by the small scope and exploratory nature of
the project. Still, there is a growing core of users and
potential users who are recognizing the advantages of the
medium and have taken an initiative in encouraging more active
participation on the part of other members and in helping to
define desirable services and features which would enhance the
value of EIES to the rehabilitation field. Many members have
also expressed their appreciation for the contact this project
has made possible with persons in other disciplines who can
contribute to their work but with whom they previously had no
available channel of communication.

In summary, the outcomes of the project to date are viewed by ISR as indicative of a need for a more interrelated user
community as well as linkages to existing data bases and the
imposition of greater structure on information which is collected and transferred via the medium. Plans for the potential future extension of the network currently include provisions for selection of a dedicated user community, linkage
to a data base resource, and the provision of on-line librar-
ian-type services to facilitate the use of the system by members and to develop specific applications. It is also anticipated that greater participation by handicapped persons can
be encouraged through the use of adaptive interface devices.

Acknowledgment

Special thanks are extended to Thomas Williams for his
analysis of interim statistical data on the project and reporting on the results to date.

8. A Semi-Visible College: Structural Effects on a Social Networks Group

Abstract

This is a report of an examination of the impact of EIES
on the structure of interpersonal relations among a set of
participants. Sociometric data were collected at the start of
participation and compared with those collected after seven
months of activity. Results suggest that EIES participation
affects not only the patterns of intellectual exchange among
participants but their more personal relationships as well.

Introduction

This is a report of the first seven months of electronic
information exchange by a collection of specialists in the
study of social networks. As a collectivity of specialists,
we are pretty much like the Futures or the General Systems
groups (as described elsewhere in this volume).   We come
from a broad range of disciplines, we are spread out through
seventeen states and two Canadian Provinces and we are involved in an emerging field of science. Like those other
groups, we are at a stage in our development as a field where
regular and significant communication is essential if we are
to develop consenses and the norms and standards that are
needed for growth.

Our overall responses to the EIES experience have also
closely paralleled those of the Futures and General Systems
groups. We succeeded in recruiting a small core of very active participants, but overall activity rates were rather low
and a number of seemingly eager applicants never appeared on

This report was funded by the National Science Foundation
(NSF-DSI77-16578). The findings and opinions expressed here
are those of the authors and do not necessarily reflect the
views of the National Science Foundation.

the system. We were able to develop a few worthwhile intellectual efforts, but most of our conferences failed to sustain
a continuing intellectual thrust; they withered and died. We
found that EIES participation greatly simplified the solution
to a number of practical organizational problems, but we were
continually frustrated by its tendency to demand attention and
intrude on our other day-to-day activities. In short, we
were--pretty much--a typical EIES trial community; we had both
positive and negative experiences.

From the beginning, we also shared with several of the
other trial groups a concern with using EIES to develop a
sense of community among the participants. In our case, however, this concern was central. As social networks scientists, we were conscious of the importance of communication in
facilitating the development of linkages between persons.
Only by establishing such interpersonal links could we hope to
arrive at the common understandings, norms and consenses
needed in a developing field.

Even before the EIES trials started, one of our number,
Mullins (1),   had suggested that the emergence of a specialty in science requires more than the sort of communication
that is provided by journals, the mails or even the telephone.
Mullins proposed that a developing field needs the kind of
communication that can engender friendship and trust--the kind
that is usually associated only with long term face-to-face
colleagueship. Thus, as networkers, one of our main concerns
all along has been a question about how well EIES could be
used to develop intimate interpersonal ties between participants.

Beginning as we did, with a sort of loosely knit collection of scholars with overlapping interests but few interpersonal ties, we wondered whether we could use EIES to develop into a real community characterized by ties of friendship and trust. Participants on EIES are at least potentially
accessible to each other. They have a sort of interpersonal
visibility that they might otherwise lack. The problem is
whether this semi-visibility can help us to evolve from a
mostly anonymous collectivity to a genuine tightly knit academy.

These were the sorts of questions that were prominent
among those asked by the social networkers as we entered the
EIES world. These questions are addressed, in a preliminary
way, in this report. We will examine the impact of the EIES
experience on the sociometry of the social networks community.

Sociometric Analysis

Data for the current analysis were generated by two
questionnaires, the first of which was administered inter-
actively on EIES at the very start of the experimental trial
period. The second set were mailed out to participants seven
months later. Altogether, 29 participants completed the
questionnaire both times. These 29 people generated the data
for the present report.

The questionnaires contained four items that are relevant
to the determination of interpersonal contacts among participants. 1. Each participant was asked to designate all other
participants that he or she had "heard of or read." 2. Participants were asked to indicate those others whom they had
"met, exchanged letters or phone calls or computer-conferenced
with." 3. Next, they were asked to indicate those others
whom they considered to be "friends." 4. Finally, they were
asked to indicate those others whom they considered to be
"close personal friends."

These are the data that are examined in the current report. They allow us to address two questions: 1. To what
extent and how were these participants linked together at the
start of the experimental trials? 2. What changes in interpersonal linkages took place in the first seven months of the
trial period?

The density of a relation is the ratio of the number of
reported ties to the total number possible. Table 1 shows the
densities for the four relations studied here. Initially,
only 62 percent of the participants were linked by having
heard of one another. And, at the same time, less than 50
percent of these pairs report having met.

Table 1. Density for four relations over two times

| | TIME | |
Relation	First	Second
heard of	.62	.77
met	.49	.68
friends	.14	.22
close friends	.05	.06

These results say something about the effects of brief
contact, since 21 of these 29 people had just attended a day-
long conference in Pennsylvania immediately before taking the
questionnaire. It supports the notion that, as Killworth and
Bernard (2)   have suggested, actual social contacts and reports of these contacts are often quite different phenomena.

The data for the second questionnaire show a considerable
amount of consistent change. There were noticeable increases
in the proportion of people reporting relationships of all
four kinds. It would seem that the computer conference, or
perhaps some other events that took place during that seven
month period, brought these people closer together.

So far, we can see that the density of ties increased
during the period of study, but we can learn more by looking
at the pattern of these ties. In order to do that, we must
examine some other characteristics of the network. First, we
shall consider how close people are to one another.

Reachability in a social network is defined as the con-
dition in which two people are able to contact one another
either directly or through intermediaries in terms of some
specified relation. Thus, if we were interested in friendship: I can reach my friend, I can reach my friend's friend,
my friend's friend's friend and so on. Moreover we can talk
of the distance along such a path in terms of the number of
links that are necessary for one person to reach another. My
friend is at distance 1 from me and his or her friend is at
distance 2.

The numbers of reachable pairs and the distances between
them in our EIES group are shown in Table 2. The number of
reachable pairs is growing wherever possible (there are 812
reachable pairs possible for these data). Moreover, the
participants are getting closer together--distances are
shrinking--on all relations but one. For close personal
friends the second questionnaire shows that, although there
are more than twice as many reachable pairs, the average distance between them is increased. Overall, then, this group
seems to be drawing together.

For close personal friends, data from the first questionnaire seemed to show the presence of tight little cliques; by
the time of the administration of the second questionnaire
there were many more personal friends reported and they were
beginning to be loosely linked together into larger structures. This suggests that at the end of the second questionnaire there was much more of a "community" among these social
networks people.

Table 2. Average distances between reachable pairs and number of reachable pairs for four relations at two times.

| | TIME | | | |
| | First | | Second | |
Relation	Distance	No.of Prs.	Distance	No.of Prs.
heard of	1.38	812	1.17	812
met	1.52	812	1.30	812
friends	2.76	728	2.18	812
close friends	2.01	96	3.13	221

We can learn something about inequalities in social choice by looking at the frequencies with which individuals were chosen for each of the relationships. Some kind of stratification is indicated by the tendency for some individuals to be chosen too often and some too seldom. An index of this tendency is given by the variance in the numbers of times individuals were selected as targets for a given relation. In general, the greater the variance, the greater the tendency for individuals to be unequal in being chosen. A large variance suggests that some people are chosen much too much and some too little. Such a tendency toward over- and under-choosing indicates that people are arranged into some kind of hierarchy. In contrast, if the variance is small, we have evidence that the relation being studied is egalitarian. Data on variances of being chosen are shown in Table 3 along with their expected values under the assumption that choices were random.

Data for the first time period show that variance in being chosen decreased markedly with increasing intimacy of relationships. This makes sense. It suggests that the tendency of people to "hear of" one another is highly stratified--participants differed in their prominence prior to their entry into EIES. On the other hand, the choice of "close personal friends" is quite egalitarian--apparently they were chosen without reference to "status."

By the time of the administration of the second questionnaire, however, things had changed considerably. The variances for the less intimate relationships had been reduced, presumably because since most people had gotten to know one

Table 3. Variances, expected variances and ratios of variance
to expectation of distributions of number of times chosen for
four relations at two times.

| | TIME | | | | | |
| | First | | | Second | | |
Relation	V	E(V)	V/E(V)	V	E(V)	V/E(V)
heard of	42.9	7.5	5.7	19.0	5.6	3.4
met	23.5	8.1	2.9	20.8	7.2	2.9
friends	9.4	3.8	2.5	20.6	5.2	3.9
close friends	1.3	1.3	.9	2.0	1.7	1.2

another, there was less effect of prior prominence. The increased variance for the more intimate relations,
however, suggest that the tendency toward establishing hierarchy has shifted from acquaintanceship to friendship. As
these people got to know one another more intimately, it
would seem, they began to stratify one another in more intimate terms. There was a shift from stratification in terms
of academic status to stratification in terms of personal
popularity.

Data in Table 4 on mutual choices can begin to tell us
something about how this stratification might work. First of
all those data show that the choices are much more predomi-
nantly mutual with increased intimacy of relationship. Less
intimate relations, then, tend not to be reciprocated, and
differences in being chosen probably reflect a stratified
hierarchy of individuals.

The predominance of mutual choices for friends and close
personal friends, however, tells a different story. For these
more intimate relations, any tendency toward stratification
probably embodies a tendency to form a hierarchy of clusters
of closely tied individuals. Thus, for friends, we seem here
to be developing a stratified set of clusters. For close personal friends, since they exhibit even more mutuality of
choice, choices probably reflect the development of an increasingly stratified set of tightly knit and unconnected
cliques.

Table 4. Number of mutual choices, expected number and ratio
of the observed number to its expectation for four relations
over two times.

	TIME					
	First			Second		
Relation	M	E(M)	M/E(M)	M	E(M)	M/E(M)
heard of	204	155.9	1.3	276	243.3	1.1
met	167	98.9	1.7	236	189.1	1.3
friends	33	8.4	3.9	53	19.3	2.7
close friends	12	.9	13.3	18	1.5	12.0

Discussion of Results

Overall, these data reveal some systematic changes in relationships among these people. They are both more aware of
one another and more willing to claim ties of friendship.
Pairs of people are closer together and fewer individuals are
isolated. There are increased numbers of mutual choices and
the patterns of stratification are switching from individual
hierarchies to hierarchies of clusters. The data suggest a
general pattern of increased density and structural organization of relationships.

These observed changes in patterns of personal ties are
particularly interesting in this context because, unlike those
observed in other longitudinal sociometric studies, the participants here had very little face-to-face contact during the
study period. Most of their contacts were computer-based.

It is difficult to tell how many of these changes are due
to the EIES experience, and how many have resulted from other
events that took place during the same time period. Anecdotal
data can throw some light on this matter. Reports from participants reveal that on at least four occasions during the
course of this project friendships were formed between pairs
of persons who had never met face-to-face. In one case a pair
of people agree that they have developed a close personal
friendship while using the EIES system.

In general, however, individual subjective reports seem
to suggest that EIES is more useful as an adjunct to personal
contact than as a substitute. For a pair of people who have
actually met, electronic communication seems to be able to
provide an adequate medium for the sort of frequent interaction that is needed in building up a friendship. Moreover,
people who are on the road to becoming friends electronically
are probably more likely to agree to attend the same face-to-
face conferences and more aware of one another if they do.
Many of our participants report a sense of "kinship" with
other EIES people when they meet at a conference.

Participants suggest that when they do meet and develop
some effectual ties they can maintain contact by using EIES.
At this point, then, the subjective daca seem to suggest that
although EIES can be used in the creation and support of more
intimate relationships, it is probably best conceived--in
this context--as an adjunct to more traditional sorts of interpersonal contacts.

Summary and Conclusions

The overall results in this study show marked changes in
patterns of linkages between the two waves of data collection.
Participants got closer together and the initial patterns of
stratification seemed to shift in form. Not only did participants "meet" via the computer, but there were noticeable increases in the extent of more intimate personal links between
them.

Taken as a whole, the results were very much like those
one might observe in a group of people brought together in a
common physical location for an extended period of time. It
is tempting, therefore, to conclude that computer-based
communication can substitute for face-to-face interaction. Of
course, such a conclusion cannot be drawn. Although this is a
comparative study, it is not experimental. The confounding
effects of history, maturation, testing and instrument decay
described by Campbell (3)   all compete with the computer experience as explanations of observed differences.

We do, however, have enough anecdotal information to suggest that the computer conference itself had an impact on the
structure of this group. It is not unreasonable, then, to end
with the conclusion that computer conferencing does seem to
provide the sort of experience that can affect patterns of interpersonal linkages. Not only can participants exchange information, but they seem to be able to exchange friendship as
well.

References

1. Mullins, Nicholas C. Theories and theory groups in contemporary american sociology. New York: Harper and Row; 1973.

2. Killworth, Peter; Bernard, H. Russell. Informant accuracy in social network data. Human Organization. 35: 269-286; 1976.

3. Campbell, Donald T. Factors relevant to the validity of experiments in social settings. Psychological Bulletin. 54: 297-312; 1957.

9. Use of Computer Conferencing to Validate and Update NLM's Hepatitis Data Base

Abstract

The contents of the National Library of Medicine's (NLM) Hepatitis Data Base, a prototype computerized information transfer system for health practitioners,are presently under- going validation and updating by consensus of a ten mem- ber group of subject experts. The Electronic Information Ex- change System (EIES) serves as the principal medium of com- munication linking the geographically-dispersed experts with one another and with the NLM staff. The computer network will also be used during successive monthly updates of the data base. This paper summarizes the experts' successful use of EIES to systematically propose changes to the Hepatitis Data Base, stemming from a thorough review of a baseline ver- sion of the data base and of approximately 850 newly-publish- ed journal articles in the field of viral hepatitis. Current use of EIES for identifying, discussing, and reaching consen- sus on controversial aspects of the data base is described.

Background

Transfer of new information into forms useful for and used by health care practitioners is a major concern and is addressed by many different approaches within the biomedical and information science communities. (1). A newly defined program area within the Lister Hill National Center for Bio- medical Communications, the R&D component of the National Library of Medicine (NLM), is aimed at the development, demonstration and evaluation of a prototype computerized in- formation transfer system in the form of a comprehensive

_____

Access to the Electronic Information Exchange System (EIES) is provided by the National Science Foundation's Division of Information Science and Technology.

literature-based knowledge bank. The purpose of the system is
to provide substantive answers to practitioners' questions,
answers that are current and are the consensus of a group of
experts, immediate responses to inquiries, data in varying
depth supporting the answers, and citations to primary publications for more detailed study if desired. The diseases
"viral hepatitis" have been selected as the initial test
model, and a data base 350 pages in length and containing
nearly two million characters, can now be explored via computer terminals at NLM. The philosophy of the system and de-
tails of its configuration are presented elsewhere. (2, 3).

Role of Computer Conferencing

     The group of ten viral hepatitis experts (researchers,
educators, and practitioners) that has been assembled to validate and update by consensus the Hepatitis Data Base, all
geographically dispersed and exceptionally busy individuals,
is linked with one another and with the staff of the NLM by
means of the Electronic Information Exchange System (EIES).
The National Science Foundation (NSF) has arranged for NLM's
trial use of this test facility.

     The objective of this research, which in effect is an
experiment within an experiment, is to determine the efficiency and effectiveness of computer conferencing as the principal mode of communication for performing the critical tasks
of data base validation and updating. With the aid of computer conferencing, it is anticipated that consensus on controversial elements of the data base can be achieved, and up-to-
date modifications effectuated in an efficient and timely
manner, and at the personal convenience of each expert without the need to leave the confines of home or office. Assuming a positive outcome of this trial with the Hepatitis Data
Base, computer conferencing will be adopted as a principal
component of the methodologies employed in constructing and
maintaining data bases in other disease areas.

     Each member of NLM's Expert Group on Viral Hepatitis,
functioning as Hepatitis Group 80 on EIES, has been provided
with a detailed set of instructions and procedures for carrying out the tasks of data base validation and updating. A
Teletype Model 43 terminal was made available by NLM for each
expert's personal use. Instructions for operating the terminal and for interfacing with EIES were specially prepared as
a truncated version of the standard EIES user manual. In the
NLM version, the use of specific EIES commands is tied directly to and interrelated with the accompanying instructions
for performing specific project tasks. Our intent has been

to provide the experts with a simplified subset of commands
and supporting documentation sufficient to perform the tasks
at hand with a minimum of training and effort. (See Table 1
for a list of these commands and their functions.) This approach was taken in recognition of the experts' extremely
busy time schedule and limited prior experience with computer
systems. However, each expert was also encouraged to explore
the full complement of EIES capabilities as time and in-
clination permitted. While this approach resulted in many
users being unaware of the more sophisticated EIES features,
it is believed that this strategy enabled us to achieve 100%
user mastery with training of the expert and/or his staff
limited to online and telephone consultations. Six of the
ten experts report that they operate the terminal themselves,
the remaining four rely on a secretary or research assistant
to execute their messaging and conferencing activities. An
eleventh member of the group serves as Chairman; his role in
the project and on EIES is qualitatively and quantitatively
different from the others, and will not be addressed here.
All members of Hepatitis Group 80 were brought online in
February, 1978; two were later substituted for reasons un-
related to their use of computer conferencing.

Of the several distinct modes of communication available
on EIES, Hepatitis Group 80 utilizes two: "messaging" which has
been used to transmit proposed changes to the data base, and
"conferencing" which at this writing is being used for discussion and resolution by consensus of controversial aspects of
the data base contents.

Messaging Data Base Changes

During a six month period beginning in March 1978 and ex-
tending thru August 1978, members of Hepatitis Group 80 reviewed and commented upon the accuracy, completeness, and
authoritativeness of a draft version of the Hepatitis Data
Base constructed by NLM staff. This task was followed by a
systematic review by the experts of approximately 850 specially selected journal articles newly published in the field of
viral hepatitis. (Journal articles are to be regularly for-
warded to the experts for review and maintenance of data base
currency.) EIES was successfully used for two- and three-way
procedural communication amongst the NLM staff, the Chairman
of the expert group, and the experts themselves. Messaging of
the text of proposed additions, deletions, and modifications
to the draft Hepatitis Data Base, typically performed by each
expert during intensive periods of activity, coincided with
the experts' ability to accomplish the assigned validation
and updating tasks in the midst of their other professional
commitments. (Preliminary system use data are available.)

Table 1. EIES Commands in NLM User Manual

Command	Command Function
++	Return to INITIAL CHOICE? point
+ON	Tells who is online
+CNM	Compose/Send New Message (or +CM)
+CY M#	Copy Message #
+GC#	Get Conference #
+CCC	Compose/Send Conference Comment
+GET M#	Get Message #
+GET C# C#	Get Conference #, Comment #
+DM# - #	Display Message Titles # through #
+DC# - #	Display Conference Comment Titles # through #
+SEARCHN	Search on Numbers/Names
+SEARCHD	Search on From-To-Dates
+SEARCHW	Search on Words/Phrases (Paragraph Numbers)
+SEARCHA	Search on Associations
+VOTEYES	Vote "Yes"
+VOTENO	Vote "No"
+ABSTAIN	Vote "Abstain"

The amount of time spent online by each of the ten hepatitis experts during the seven month period beginning February 1978 and ending August 1978 (i.e., one month start-up time and six months devoted to the actual performance of the data base validation and updating tasks) is shown in Table 2. A total of 194 online hours were used, ranging from a low of 4.5 hours expended by one member to a very active 66 hours recorded by another; the median is approximately 16 hours.

It should be noted that EIES usage did not end subsequent to the conclusion of these tasks in August, but dropped to a much lower level of activity pending initiation in January 1979 of the next project task requiring communication on EIES. Usage by NLM staff, while not reported here, continued at a moderately high level.

Table 2. Hepatitis Experts' Online Hours

EXPERT	MONTH							TOTAL
	FEB	MAR	APR	MAY	JUN	JUL	AUG	
A	1:15	4:45	1:11	8:10	1:25	0:13	3:23	20:22
B	0:24	1:26	4:13	1:15	0:08	0:39	0:29	8:34
C	0:03	1:31	0:35	0:15	0:17	1:25	0:20	4:26
D	1:26	9:24	4:21	0:46	0:34	1:22	0:13	18:06
E*	--	--	--	--	--	3:49	6:15	10:04
F	3:13	4:31	8:06	28:14	2:02	6:45	13:35	66:26
G	0:02	--	--	0:58	0:07	6:37	18:06	25:50
H*	--	--	--	--	--	3:54	3:28	7:22
I	0:02	--	9:17	2:24	0:01	3:55	0:13	15:42
J	0:24	0:48	12:53	1:40	0:07	1:27	--	17:19
TOTAL	6:49	22:25	40:36	43:32	4:41	30:06	46:02	194:11

(*) Joined Hepatitis Group 80 in July 1978.

Table 3. Number of Messages (Lines) Sent by Experts

EXPERT	MONTH							
	FEB	MAR	APR	MAY	JUN	JUL	AUG	TOTAL
A	2 (8)	4 (44)	4 (16)	7 (251)	--	--	4 (96)	21 (415)
B	2 (13)	7 (121)	48 (613)	14 (191)	1 (2)	6 (80)	2 (59)	80 (1079)
C	--	2 (106)	--	--	--	1 (46)	--	3 (152)
D	1 (5)	18 (236)	10 (121)	1 (10)	3 (121)	7 (56)	1 (3)	41 (443)
E*	--	--	--	--	--	4 (15)	16 (109)	20 (124)
F	2 (12)	3 (27)	4 (112)	24 (757)	--	4 (172)	17 (705)	54 (1785)
G	--	--	--	1 (4)	--	16 (517)	34 (1158)	51 (1679)
H*	--	--	--	--	--	7 (154)	8 (47)	15 (201)
I	--	--	25 (745)	4 (87)	--	15 (300)	--	44 (1132)
J	1 (1)	--	21 (543)	3 (53)	--	4 (63)	--	29 (660)
TOTAL	8 (39)	34 (534)	112 (2150)	54 (1353)	4 (14)	64 (1403)	82 (2177)	358 (7670)

(*) Joined Hepatitis Group 80 in July 1978.

The number of messages sent during this period by the individual experts for purposes of validating and updating the Hepatitis Data Base is shown in Table 3. Because the variability in message length across experts is substantial, the number of text lines comprising the message content is also presented for comparison. A total of 358 separate procedural and substantive message items, containing more then 7600 lines of text, were transmitted to NLM. These ranged from a high of 80 messages (or 1800 lines) to a low of only 3 messages. The content of the messages dealt with proposed substantive and organizational changes to the data base stemming from the experts' review of the initial draft version of the data base and of the selected journal articles. These changes were subsequently incorporated in the data base on a tentative basis pending ratification by the group as a whole. (See section on "Conferencing and Consensus Development.")

It is difficult to draw unambiguous conclusions from these preliminary data. While the amount of time spent on-line appears to be roughly correlated with the number of messages sent and/or number of text lines transmitted, there is probably a good deal of individual variability in the speed with which messages are composed. (Note that these tallies also include time spent receiving messages, which at present cannot be separately broken out.) So, too, are there characteristic and consistent differences observable in message length across the individual experts, and the number of proposed data base changes actually encoded in a given message item. Experience seems to indicate, however, that messaging on EIES can be a convenient, quick, and accurate means for transmitting complex biomedical data from multiple participants simultaneously (and from persons who are not "computer-wise"), and for effectively managing the information flow for subsequent retrieval, editing, and insertion in the data base. This conclusion should perhaps be tempered by the reality of the not insignificant number of system crashes, attributable to both EIES and TELENET malfunctions, that occurred during this phase of the trial. Happily, these disturbances appear to have greatly dimininshed in number.

Conferencing and Consensus Development

A second draft of the Hepatitis Data Base, incorporating the substantive and organizational changes proposed by each expert, has been constructed and was mailed to the experts in January 1979 for their review and ratification. Until now, the characteristic communication behavior of Group 80 members has been to log-on the system, message their proposed data base changes (addressed to NLM staff and the group Chairman), and log-off. Little social or non-task-directed communication

appears to have taken place thus far, despite the fact that
each of the experts is known to the others and nearly all
share membership in a common invisible college. Communication
of a more personal nature may be expected to increase during
the current phase of the trial as controversial issues per-
taining to data base content are identified, discussed and re-
solved by consensus within the framework of Group Conference
80. Procedures for structuring these consensus development
discussions, and bringing the issues to a vote, have been established along the lines of a modified Delphi technique. As
before, specially written and programmed EIES instructions are
tied directly to the assigned task procedures, and are meant
to be used in a complementary manner.

At the conclusion of the consensus development phase,
which is projected to run approximately three months, each expert and his staff will be queried in depth by means of a
structured questionnaire and personal interview as to their
perceptions and experiences concerning the extent to which
EIES facilitates or inhibits the performance of their assigned
tasks. The preliminary system use data reported above will be
expanded upon and subjected to analysis in light of these
findings. Cost-benefit analyses will also be performed as
part of an overall evaluation of the methodologies and procedures used to construct, maintain, and operate as a service
the Hepatitis Data Base prototype. The development of additional data bases, each requiring the collaboration of groups
of subject matter experts, is now being planned. It is anticipated that these evaluation data will enable us to determine
the utility and acceptability of computer conferencing as a
key component of the NLM's Information Transfer Program.

Acknowledgments

The author gratefully acknowledges the contributions of
Lionel M. Bernstein, MD, Director of NLM's Information Trans-
fer Program and "text composer" of the draft Hepatitis Data
Base, and Mrs. Frances Peneburgh who organized and processed
the voluminous output of NLM's computer terminals. Members of
the Expert Group on Viral Hepatitis are Hyman J. Zimmerman,
MD, Veterans Administration Hospital, Washington, Chairman;
John A. Bryan, MD, Center for Disease Control, Atlanta,
Georgia; Thomas C. Chalmers, MD, Mt. Sinai School of Medicine,
New York, NY; Burton Combes, MD, University of Texas South-
western Medical School, Dallas, Texas; F. Blaine Hollinger,
MD, Baylor College of Medicine, Houston, Texas; Jay Hoofnagle,
MD, National Institutes of Health, Bethesda, Maryland;
Erwin M. Kammerling, MD, private practice, Chicago, Illinois;
Raymond S. Koff, MD, Veterans Administration Hospital, Boston,
Mass.; James W. Mosley, MD, University of Southern California

School of Medicine, Los Angeles, Calif.; Leonard B. Seeff, MD, Veterans Administration Hospital, Washington; Wolf Szmuness, MD, New York Blood Center, New York, NY.

References

1. Schoolman, Harold M.; Bernstein, Lionel M. Computer use in diagnosis, prognosis, and therapy. Science. 300: 926-931; 1978 May 26.

2. Bernstein, Lionel M.; Siegel, Elliot R.: Ford, William H. The hepatitis knowledge base prototype. Proceedings of the IEEE second annual symposium on computer application in medical care. Washington, DC; 1978 November.

3. Siegel, Elliot R.: Information transfer for health practitioners: an innovative approach. Proceedings of the 41st annual meeting of the American Society for Information Science, New York, NY; 1978 November.

Peter Johnson-Lenz, Trudy Johnson-Lenz,
J. F. Hessman

10. JEDEC/EIES
Computer Conferencing for
Standardization Activities

Abstract

The Joint Electron Device Engineering Council (JEDEC),
under the aegis of the Electronic Industries Association
(EIA), is experimenting with the use of electronic information
exchange in selected aspects of its work of promoting hardware
and software standardization in microprocessor/large scale
integration (LSI) products. Six potential advantages in using
computerized conferencing for JEDEC committee and task group
work are described. JEDEC's activities using the Electronic
Information Exchange System (EIES) are discussed, including
the use of specially tailored software to assist members in
collective development of terms and definitions, a first step
in setting standards. Automated data collection, as well as
evaluation and facilitation of the project, are described.
Several future enhancements, including the use of intelligent
terminals, are mentioned. The JEDEC-EIES project is one of
the NSF-sponsored operational trials of electronic information
exchange for small research communities.

Introduction

The Joint Electron Device Engineering Council (JEDEC) began its operational trial use of the Electronic Information
Exchange System (EIES) in the fall of 1978. JEDEC is an
association of firms concerned with the manufacture of solid
state electronic components under the aegis of the Electronic
Industries Association (EIA).

This project is supported by grant NSF-DSI 78-09189 from the
National Science Foundation. The opinions and findings reported here are solely those of the authors and do not necessarily represent those of the National Science Foundation.

JEDEC's use of EIES involves "selected aspects of its
work of promoting hardware and software standardization in
microcomputer/large scale integration (LSI) products, with
special attention to microprocessors" (proposal to NSF under
program announcement NSF 76-45).

JEDEC's standardization activities are conducted by a
series of numbered committees (e.g., JC-42 Committee on Semiconductor Memories) and decimal numbered task groups within
the larger committees (e.g., JC-42.1 on Bipolar Memories).
The numbered committees generally meet quarterly in various
locations around the U.S. According to preliminary evaluation
data, the average cost of attending these meetings is a little
over $400 per person, excluding time lost from work. The
meetings generally last 1-3 days. There is some communication
between meetings among committee or task group members, generally by phone or by phone and mail. Turnover and absentee-
ism are problems with these face-to-face quarterly meetings.

The JEDEC-EIES project is designed to see if some of the
committee or task group work on standardization can take place
via electronic information exchange and to see how this
relates to the current practice of face-to-face meetings and
to (mail) letter balloting on standards.

Potential Advantages of
Computerized Conferencing for
JEDEC Committee Work

Computerized conferencing offers several potential advantages to user groups such as JEDEC committees and task
groups. First, it is asynchronous; that is, people can enter
and retrieve material from the system at their own convenience, without having to be on line at the same time. Much
between-meeting JEDEC work is done on the telephone, and members complain about the difficulty of arranging such phone
conversations, given different time zones, meetings, and travel schedules. They also complain about the difficulty of
conducting conference calls involving more than a few people.

In many of the quarterly face-to-face meetings, JEDEC
members must defer making comments until they can go back to
their companies and confer with technical people or management
about proposed standards. With on-going work being conducted
asynchronously via computerized conferencing, such consulta-
tion can take place in the normal flow of business, without
having to put off standardization decisions from meeting to
meeting because the right technical people have not been con-
sulted.

Second, computerized conferencing provides a complete
written transcript of what has been discussed. This allows
any "latecomers" to catch up in a discussion without holding
up the rest of the conversation. With the moderate turnover
in JEDEC committees and task groups, this is particularly
important. Furthermore, the transcript can help eliminate
confusions about what was decided at a given meeting. As long
as there is any confusion about what occurred, any member can
continue to push his own interpretations of a standard at successive meetings, thus delaying any general agreement for
months and months.

Third, computerized conferencing can help speed up the
process of meeting and coming to decisions. At the moment, it
can take up to two years for a standard to be set through regular face-to-face meetings. With the dynamic nature of the
semiconductor industry, a given component may be obsolete by
the time a standard is set for it. One task group chairman
spoke wistfully of a quick iteration in JEDEC decision making
taking only three months! In contrast, preparation for an
EIES demonstration for the meeting at which this group chairman made his comment was conducted via EIES over a two-week
period, with several iterations to figure out the appropriate
focus for the demonstration. In the two weeks, twenty-seven
conference comments were generated, including items made into
viewgraphs for a committee presentation. Several people in
different locations were involved in this planning conference.

Fourth, computerized conferencing provides facilities for
both private messages and for on-going conferences involving
many people. Communication among committee and task group
members can take whatever form is appropriate. Delicate,
classified, or proprietary material can be discussed among individuals with complete security, while general conversations
can take place in semi-public forums.

Fifth, a computerized conferencing system such as EIES
also offers rudimentary graphics capabilities. This is very
important to groups such as JEDEC which deal with graphic
representations of various products. After some experimenta-
tion, it seems that many of the graphics JEDEC circulates can
be transmitted through EIES, with the exception of detailed
schematics. These will be circulated through the mails and
then referred to in on-going conferences on EIES.

Sixth, and perhaps most important, EIES is a programmable
system. Its high-level programming language, INTERACT, provides the capacity for creating tailored software to help
groups in their decision-making processes. Several such programs have been written for the JEDEC project.

JEDEC's Use of EIES

JEDEC is actually composed of a number of committees and
task groups. Because computerized conferencing is a new communications medium and most JEDEC members are very busy people, it has taken awhile to get various members to experiment
with using EIES. Despite problems in recruiting JEDEC members
to make regular use of EIES, there are now two task groups
which are beginning to do their work via computerized conferencing. One is JC-13.4, a task group involved in revising a
military specifications document on microcircuits to include
material on microprocessors. The military had to give permission for EIES to be used for this activity. JC-13 is a
committee on liaison to the military. The second group,
JC-42.11/JC-11.3.1 is a newly formed task group that is serving as an interface between one group concerned with memory
packaging standardization and another with chip carrier packaging standardization. It includes members from JC-42 on
semiconductor memories and JC-11 on mechanical standards. All
members of these task groups using EIES are in different companies and geographic locations.

A great deal of JEDEC's work involves agreeing on a set
of terms and definitions that apply to a given standardization
topic. After terms and definitions are agreed upon, para-
meters are then worked out, such as electrical voltages, tem-
perature tolerances, testing methods, and so on. To make a
group's work easier in developing a set of terms and definitions (actually, a glossary), a special software system has
been written in INTERACT. Called +TERMS, this system allows
any member to add a term (abbreviation and full name), add a
definition, make a comment about a term or definition or another comment, and vote on terms and definitions. All new
terms, definitions, and/or comments are put into a pending
items "queue," just like waiting messages and conference comments within regular EIES. Whenever a member enters a given
glossary within the +TERMS system, he/she is told how many
waiting terms, definitions, and comments there are and is
given a chance to accept them. Thus, each glossary within
+TERMS is like a terms and definitions conference, except that
there are different kinds of items within the conference
(terms, definitions, and comments) as well as special software
commands to manipulate these items. Members may vote on pre-
ferred terms and definitions by entering simple commands.
Straw votes may be entered anonymously. Tallies of the votes
are available to members. There are currently three JEDEC
glossaries in development on EIES, and a fourth is planned.

After more JEDEC members have had experience in using the
+TERMS system, and glossary work is almost done, a similar

system may be written for collective development of para-
meters.

Evaluation and Automated
Data Collection

As with other groups using EIES as part of NSF's Operational Trials Program, the JEDEC project has its own evaluators who are assessing its use of computerized conferencing
during the period of the grant. Pre-use questionnaires were
developed which asked about JEDEC members' reasons for participation, expectations of using EIES and its relationship to
the regular JEDEC face-to-face meetings, experience of quarterly face-to-face meetings, participation in standardization
activities, communications skills, experience with and access
to computer terminals, and demographics. These questionnaires were sent to all JEDEC-EIES members through the mail,
with instructions that they could fill out the questionnaires
and mail them back, or they could answer the same questions
during an on-line automated interview. To date, about half
of the respondents have returned the questionnaires through
the mail and half have filled them out on-line. Giving members this option has increased the response rate to the pre-
use questionnaire to a virtual 100%.

Formal letters have been sent to all participants via
EIES asking for permission to collect information about whom
they communicate with on EIES during the course of the project. This information will allow the evaluators to see if
communication increases or decreases during the period of the
operational trial, as well as to monitor any committee "cross-
talk" that may develop. After all the permissions are gath-
ered, again via EIES, the information about who communicated
with whom and how many lines of text were sent and received
will be recorded automatically. No message content will be
looked at. Complete confidentiality of the data will be maintained.

Facilitation

This project also includes a facilitation team to help
JEDEC participants make the best use of EIES. Having no prior
knowledge of JEDEC standardization activities, the facilitators are encouraging all JEDEC-EIES members to participate in
planning and design discussions about various aspects of the
project. Conferences have been opened for discussions of
software design, evaluation design, demonstration planning,
and task group activity planning. The advice, suggestions,
and expressed needs of participants have been invaluable in
project facilitation to date.

Looking Toward the Future

Several JEDEC members have been experimenting with using intelligent terminals so they can compose off line and then transmit to EIES in "batch." This is particularly useful when members need to circulate documents for comment, rather than shorter text items. In addition, it saves connect time. One member has also defined a series of EIES interactions under one command name so that his secretary can log on the system and retrieve all waiting items for him (messages, conference comments, etc.) and then log off -- with only one command.

Along with the beginning use of intelligent terminals, the JEDEC project may also provide an opportunity for experimenting with intelligent graphics terminals.

While it is much too early in the JEDEC-EIES project to draw any definitive conclusions, JEDEC does seem like an excellent candidate for use of computerized conferencing for on-going committee and task group work. JEDEC members on EIES are generally middle- or upper-level managers in the companies they represent, and they have a moderate to strong motivation to meet and confer with other committee members in order to set standards for their products. They are generally technically "initiated" if not sophisticated; all members have used computer terminals before, if only for playing games on a computer. They already have a regular committee and task group structure, and they participate in on-going task-oriented activities--the various aspects involved in setting standards.

The perceived advantages of asynchronicity, written record, fast interaction, different types of communication (messages, conferences, glossaries), graphics, and special tailored (software) communications structures make computerized conferencing very attractive for JEDEC committee and task group work. This project is still in its early stages. It will be interesting to see if the potential of computerized conferencing for enhancing JEDEC's standardization activities is realized.

Peter Johnson-Lenz, Trudy Johnson-Lenz

11. LegiTech/EIES: Information Exchange Among State Legislative Researchers

Abstract

The LegiTech-EIES project is one of the operational
trials of electronic information exchange supported by the
National Science Foundation. Legislative researchers in 25
states and resource-reviewers in a like number of Federal
agencies and other resource organizations are using EIES to
exchange inquiries, responses, and leads about scientific and
technical matters of interest to state legislatures. Specially tailored software has been developed within EIES to provide
LegiTech participants with computer-supported entry of inquiries, responses, and leads; with the ability to select only
those inquiries of interest; and with automated data base
management of all the information. This software is described
briefly. The relationship between LegiTech and the National
Conference of State Legislatures' (NCSL's) Legislative Information System is discussed. A description of types of information needed by legislatures leads into a general discussion
of information exchange processes for policy makers at other
levels of government through a series of "PoliTechs." Further
enhancements using intelligent terminals and microprocessor-
based interconnections of computer systems are mentioned.

Introduction

"Have any states tried the 'energy stamp' approach to
assist low income persons in paying their energy bills?"

This project is supported by grants DSI 77-17838 and
DSI 77-27940 from the National Science Foundation. The
opinions and findings reported here are solely those of the
authors and do not necessarily represent those of the
National Science Foundation.

"What are the alternatives for disposal of sludge?"

"Has your state passed any laws restricting the construction of nuclear and fossil fuel plants?"

Science and technology researchers serving state legislatures have recently posed these and other inquiries to their counterparts in other states through the LegiTech computerized conferencing experiment. The responses and leads the researchers receive to their inquiries help create a collective knowledge base of scientific and technical information that will be available to all states participating in the project.

Coordinated by Participation Systems Inc. (PSI), LegiTech is an interstate network funded by the National Science Foundation to document, develop, and test various methods of exchanging technical information among legislatures and the science community. Two major types of information are being exchanged: "factual" inquiries and responses, including background information and technical resources; and "process options" which describe the purpose, features, and requirements of a variety of techniques and systems which legislatures can use to obtain technical information on a variety of issues. Dr. Chandler Harrison Stevens, president of PSI, is the principal investigator for the LegiTech project. He is being assisted by Dr. Valarie Lamont and regional nodes in the network acting in a liaison capacity.

The LegiTech Network

Since 1977, three states, Massachusetts, Minnesota, and Pennsylvania, have been exchanging technical inquiry/response briefs through the mails. These briefs are generally no longer than two pages, and they contain a short title, the question being posed (no more than three lines), background information about the question, and whatever responses and/or leads to further information have been located by the legislative researcher. These completed inquiry/response briefs have been made available to other interested state legislative researchers through the National Conference of State Legislatures (NCSL), which distributes an order form for these briefs in its monthly newsletter, "Science and Technology for the Legislatures." A similar order form is also being included in the monthly newsletter of the Congressional Clearinghouse on the Future, "What's Next?". A notebook of "process options" will be published soon and distributed by PSI and NCSL.

As the next step in the development of the LegiTech network, the National Science Foundation has funded the use of computerized conferencing by state legislative researchers as part of the NSF Operational Trials of Electronic Information Exchange for small research communities.

The LegiTech-EIES Project

The LegiTech-EIES project began in the fall of 1978. Several months were devoted to initial project organization and to the design and development of specially tailored software to support the information exchange among state legislative researchers. The researchers themselves began to use EIES in earnest in January, 1979, as the legislative sessions began. Currently researchers in half the states are members of LegiTech-EIES. In addition, there are a number of resource-reviewers located in Federal agencies, the Executive Office of the President, and other organizations, who are also participating in the legislative inquiry/response process.

Pioneering legislative researchers from Massachusetts, Minnesota, Pennsylvania, and Illinois began using EIES in November 1978 on an experimental basis to work out any problems before the rest of the states joined in. To date, about 194 inquiries have been entered into the system as well as 332 responses and/or leads to further information. Most of these inquiries and responses have appeared within the last two months, as more researchers begin to use the system.

Edited, completed inquiry/responses have also been entered into EIES in a public conference where they are available to everyone. To date there are thirty-four completed inquiry/responses on topics ranging from the use of charcoal as an energy extender to the levels of nitrites in meats and vegetables.

In addition to this exchange process, legislative researchers and LegiTech resource-reviewers may also participate in several conferences of interest. There are several organizational conferences in which project policy matters are being discussed. There are also "spin-off" conferences for more indepth discussions of technical issues. For example, NCSL's MISTIC (Model Interstate Science and Technology Clearinghouse) coordinator is starting a conference on toxic/hazardous substances. The conference will cover a review of Federal legislation on this topic, past state actions considered or on the books, policy considerations from the states' perspectives, existing intergovernmental coordination/cooperation, and public participation requirements.

Since computerized conferencing is asynchronous (that is, each participant can enter and retrieve material at his/her own convenience), this is an excellent medium for geographically dispersed legislative researchers to use. Prior to the use of EIES, researchers had to rely on the mails or the telephone to ask researchers in other states about these matters. With legislative researchers from Maine to Hawaii participating, it would be difficult to coordinate telephone calls over five time zones, but computerized conferencing makes the exchange much more convenient. Since these researchers are scattered all over the country, the LegiTech project also has a liaison in each region to assist the researchers in getting on line and learning to use the system productively. This decentralized liaison activity provides for training and assistance in a more personal, localized way and has been quite successful in getting widespread participation to date.

Special LegiTech Software

Legislative researchers are very busy people who have a need to gather pertinent information for their legislatures and who do not want to be bothered with what they consider to be irrelevant questions. The general conferencing features of EIES provide a space for relatively unstructured exchange and discussion, which is useful for brainstorming, general discussions, and extended conversations on topics of interest. However, general conferencing does not make it easy to find just the facts of interest without having to read all the rest. To facilitate the information exchange among state legislative researchers, specially tailored software was developed for the LegiTech project which is actually a subsystem within EIES.

When a researcher enters the special LegiTech system, he/she is told how many new inquiries, responses, and leads are waiting. LegiTech inquiries are posed in a special format: short title (up to 25 characters), question (up to 3 lines), and background information (up to 50 lines). As each inquiry is printed out, the researcher receives only the title and the question initially. This brief information is enough to let the researcher determine whether he or she is interested in that particular inquiry. For example, a researcher in Nebraska may not care about an inquiry relating to marine fisheries policies. If the inquiry is of interest, the researcher then "selects" it, which means that the background information is delivered, along with any responses and/or leads entered to date. Then, for each inquiry so selected, the researcher will also receive any new responses and/or leads entered later by other researchers or resource/reviewers. These new responses and/or leads will be automatically delivered. Furthermore, if a researcher does not select a given inquiry initially, he or she can always go back later and select it. Thus, the LegiTech software acts as a "filter," insulating the researcher from information or questions that are irrelevant. This is an important point for legislative research units which may receive so many inquiries through the mail or on the telephone that they can't get regular work accomplished!

The LegiTech system has been designed to be highly similar to the regular EIES structure so that researchers and resource-reviewers do not have to learn two very different systems. Users of the system specify what they want to do by entering choices from numeric menus. The system guides them through the process of composing inquiries, responses, and leads, all in the specific formats that have been developed. There is also software which allows any LegiTech member to review the activities of other members--to see who is actively using the system and which inquiries they have selected as being of interest.

LegiTech's Relationship to NCSL's Legislative Information System

The National Conference of State Legislatures has a computerized Legislative Information System (LIS), which is also funded by the National Science Foundation. This system grew out of experiences in running a research clearinghouse on an experimental basis. The fully operational LIS provides an information and retrieval capability to all state legislatures for accessing finished research reports. These reports are sent to the LIS staff where they are abstracted and entered into the LIS computer in Washington, D.C. The information in the computer includes the abstract, the name of the state which produced the report, and a listing of other states mentioned in the report. About 40 percent of the research reports prepared by legislative research bureaus in any one state deal with policy impacts or situations common to other states as well. State legislative researchers can access the information in LIS directly, or they can make requests of the LIS staff. This is an important information resource for legislative researchers.

In contrast, the LegiTech system deals with research in progress, rather than finished research reports. As such, it is a complement to (and in some cases a source of information for) the Legislative Information System. In LegiTech researchers are exchanging information about topics of interest in an attempt to pool their collective knowledge about an issue. Once this exchange process has taken place, often over several weeks with many people being involved, a given legislative researcher may take all the information and synthesize it into a research report which then may be entered into LIS. The synthesized information may also be distributed to other legislative research units via LegiTech as an edited inquiry/response. Thus, both systems are designed to help state legislatures share science and technology research information in a variety of formats. NCSL is participating in both projects.

Legislative Information Needs

Fundamentally, two kinds of information are useful to people: information that fills a gap in existing knowledge and information that expands the scope of that knowledge in an unanticipated or surprising way. Legislatures and other policy makers need both kinds of information, and the LegiTech system is designed to deal eventually with both.

Gaps in existing knowledge are relatively easy to identify and these become questions. Thus a legislative researcher may be asked by a legislator, "What are some methods of forecasting peak load electric demand?", because the legislator is not sure he or she should trust a utility's forecasts. If the researcher does not know the answer, a pointed question can be posed to other researchers within LegiTech (as well as other sources within the researcher's personal network of information resources). In this case, the researcher is looking for information to fill an existing gap in his or her knowledge. The inquiry/response format used by the LegiTech network addresses just this sort of information in a brief, straightforward form.

In contrast, one cannot ask about something which is not within one's ken. "Surprise" or unsolicited information which opens up one's understanding to include new things is also very important. Such information might be in the form of suggestions or models or novel ideas. The LegiTech project is beginning to supply this second type of information in the form of "process options" which describe generic techniques, methods, or models that may help improve legislative processes and, in particular, legislative research processes. In time these "options" may be expanded to include "policy options" as well, which will deal with alternative ways of looking at policy questions.

Inquiries and responses are entered into the system in response to specific legislative questions. Options are unsolicited, unrequested information that may be of interest to legislative researchers and legislatures. Such options are researched and prepared by the LegiTech liaison staff and others who happen to discover "promising practices" that should be shared with legislatures.

Because these options contain unsolicited information, it is important that they be structured and formatted in a manner that does not overload the legislative researchers who receive them. Each option contains a short title, purpose (not more than 3 lines), overview, features, requirements, and implementation steps. Like the inquiry/responses, the options are formatted to be no longer than 2 pages. The option overview parallels the inquiry background, and the features, requirements, and implementation sections parallel the response and/or leads in an inquiry/response. As in receiving inquiry/responses, a researcher can look at only the title and purpose of an option and then decide whether to look at the rest.

The LegiTech-EIES project has begun with the exchange of inquiries and responses, since they are easy to understand and they fill a current need for legislative researchers. The capability of handling "process options" is being discussed for inclusion in the special LegiTech software. In the meantime, these "process options" are being distributed as part of the LegiTech NetWorkbook developed by PSI, with the help of an editorial board and oversight committee drawn from NSF, NCSL, the Congressional Research Service, and several state legislatures.

The Future of Exchange Networks
for Policy Makers

The EIES computerized conferencing system includes a special high-level programming language called INTERACT in which various kinds of human communications structures can be created within EIES. The current LegiTech software is a good example of a particular structure within EIES that is separate from but similar to the other conferencing features. This software was created to assist researchers in entering their inquiries and responses within an existing format shared by several states, to filter out information not of interest to each researcher, and to keep track automatically of all the LegiTech information in a highly structured data base.

The features of this special LegiTech software can be extended and made more general to support the information exchange needs of other groups. For example, researchers in the state executive branches could use a similar system as well as Federal agencies (intra-agency and inter-agency exchange), cities, neighborhoods, professional societies, and others. Some design discussions have begun about the development of

"PoliTechs" -- a whole series of exchanges for policy makers
in various levels of government. A general system could support a number of separate exchange processes with the possibility of cross-fertilization and synergy among the exchanges.
The norms for such sharing and cross-fertilization would have
to be worked out in a manner that would maintain the separation of various branches of government and the confidentiality
of research information while still making possible extensive
sharing and networking when and where appropriate.

Another interesting future possibility includes the use
of intelligent terminals. These terminals have their own limited memory capacity so that information can be entered into
them off-line and stored. Then the stored information can be
rapidly transmitted to a system like EIES for sharing with
others. In addition, use of an intelligent terminal simpli-
fies the process of using a number of different computers for
data base searches, conferencing, or other purposes. Access
and search protocols can be entered in the terminal's memory
under simple, convenient command names so that researchers do
not have to become technically proficient in the use of several computer systems.

Furthermore, EIES includes a microprocessor (called
HAL ZILOG) which can be programmed to dial up and search remote data bases, such as LIS, the Congressional Research Service's SCORPIO system, and so forth. Once this microprocessor is programmed to handle the search protocols, access, and
billing arrangements, EIES becomes an entry point to many resources of interest to legislative researchers, or for that
matter to any policy maker. The combination of access to resources and structured communications capabilities for information exchange makes a system like EIES very promising for
meeting the information needs of policy makers on all levels.

Conclusion

The LegiTech-EIES project is an outgrowth and extension
of some limited information networking among state legislative
researchers begun in 1977. This project uses the medium of
computerized conferencing to support the exchange of scientific and technical inquiries and responses of interest to state
legislatures. Special software has been developed to help researchers serving legislative bodies in 25 states enter and
retrieve only those inquiries, responses, and leads to further
information of particular interest to them in a brief, yet
comprehensive format. In addition to the legislative researchers, there are also LegiTech resource-reviewers representing Federal agencies such as the Environmental Protection
Agency, regional bodies such as the New England Legislative

Leaders Energy Project, and other government resources such as
the White House Information Center. Regional liaisons provide
training and technical support for participants in the Legi-
Tech project on EIES. Information currently being exchanged
through EIES includes inquiries, responses, and/or leads on a
variety of scientific and technical policy issues, but the
capabilities will be expanded to include "process options" as
well. This type of electronic information exchange can be
generalized to serve policy makers at a number of levels of
government in a series of separate but overlapping "Poli-
Techs." The use of intelligent terminals and the microprocessor capability within EIES can further enhance policy
makers' access to information about factual matters and about
new approaches to the process of policy making, as well as new
ways of thinking about old (and new) problems.

12. The Designer's View

Introduction

The design, operation and evolution of the Electronic Information Exchange System (EIES) is based upon a number of observations at both a general and specific level. At the philosophical level, the process of design of the person-machine interface for interactive computer systems to service heterogeneous populations is still very much an art form. We have much in the way of lore on how to design but very few proven hypotheses.

At the level of application, the use of computers as a device to facilitate the human communication process is still in its infancy. We are now beginning to realize that when we design a communication structure to operate within an interactive computer system for a group of humans, what we are really designing is a human system. It is an electronic social system where the properties or behavior of the group are a result of an inseparable combination of human psychology, group sociology and the characteristics of the design.

At the task specific level, such as the design of a system to support scientific and technical information exchange prior to publication, there also exists a high degree of uncertainty or alternative hypotheses about the behavior of scientific groups and their underlying motivations.

The work reported in this paper is partially supported by the Division of Information Science and Technology (DSI77-21008) and the Division of Mathematical and Computer Science (MCS78-00519) of the National Science Foundation. The opinions expressed in this paper are soley those of the author and do not necessarily represent the views of the National Science Foundation.

In addition to the unknowns, a few facts could be counted upon when EIES was being laid out on the drawing boards. The user population would come from a diversity of organizations, they would range in experience from never having used a terminal or computer to being very sophisticated with respect to the use of computers. Some would interact with the system directly, while others would use intermediaries such as students, staff or secretaries. There would be no mechanism for providing direct training. Users would have to learn from what they were sent in written form and what they could gain on the terminal itself. There would be no way to impose any standardization on the behavior of users or the types of terminals to be used as is possible when the user community is part of one organization.

Design Considerations

Given the above predictables, how in fact does one go about designing a system such as EIES? The resulting conclusion is quite simple in concept, but very difficult in execution. One does not attempt to design a system as a template to fit a carefully defined set of human needs against a highly structured set of technological capabilities. Instead one attempts to design a process able to capture human needs in time and to adjust the capabilities of the system to respond and adapt to those needs. In the jargon of the social sciences, EIES is designed as a "stimulus-response" process where neither the stimulus nor the response can be held fixed in time or category. The humans as individuals and groups provide the stimulus and the capabilities of EIES are the response. What we do know about such systems where neither stimulus nor response can be held constant over time is that there is no single correct methodology or instrumentation that can be applied to their study. As a corollary to this, EIES is perhaps one of the most studied and observed of any interactive system that has ever been designed. The evaluation process itself is also an unconstrained "stimulus-response" system in the hope that some complimentary variety of evaluation methodologies will prove to be an appropriate instrumentation of such systems. Since the design must also be responsive to this need by a select subset of the EIES user community, the problem for the designer is compounded considerably. We are simultaneously exploring the design of such systems and the methodologies of evaluating them.

In light of the above, perhaps the best way to explain what is meant by designing a process instead of a system, or designing a system as a process (depending on one's philosophical view), is to delineate the characteristics of EIES which make its evolution a conscious process.

Characteristics of EIES

One advantage EIES has over other interactive systems is that it is primarily a communication system. Therefore, the human involvement in the process can be made an integral part of the system. In terms of EIES, this means that the design and implementation group, the user consultants, the evaluators and the user community at large form elements of both a formal and informal communication network governing the evolution of the system. As independent groups they each have their missions and independent communications channels such as task-oriented conferences and messages.

The user consultant group is perhaps key to this evolutionary process. These individuals were chosen not only for the purpose of aiding people in the mechanics of the system but also for their ability to act as facilitators for communications among humans. They form the nucleus or glue for the user community. Successful interactive systems always seem to include an atmosphere among the user community of mutual aid and mutual exchange of information about using the system. The user consultants act to facilitate this exchange at two levels, one on the mechanics of the system and one on the human process, such as the norms of this new mechanism of communication. The evaluation process to date seems to have established that the user consultants systematically gather and collect for analysis by the system evaluator all of the questions to which they respond. They also synthesize needs into suggested requirements for the design and implementation team, and compose documentation on new features as they are generated. As volunteers, they have no organizational biases towards the system. Rather their loyalty is to the concept of this form of communications as a new medium worthy of their investment of time in terms of the potentials it holds for themselves and others.

The design and implementation team, because of their knowledge of computers and of the capabilities of the technology, have the responsibility of reinterpreting requirements from the short-term perspective the user usually sees to the long-term generalization of that requirement. In other words, a specific desire by a user to correct a single deficiency in having the system satisfy some current need may only be a subset of a wider requirement that may be more beneficial to the user community as a whole. To incorporate a number of single specifics as changes may box in the technology with respect to later gaining the greater utility of a more comprehensive design change. It is this observation that begins to expose the difficulties inherent in the process. A parallel exists for the relationship of the user consultants to the evaluators.

Humans or groups are not always able to make explicit the underlying cause of their behavior. Ability to conceptualize design changes is limited not only by lack of understanding of the technology but also by lack of ability to relate to fundamental psychological and sociological factors. Since we are dealing with the uniting of a technological system and a human system, we need the complement of the technological designers on the human side--that is the role that the evaluators fill. Theirs is the job of correlating information from numerous sources, including the user consultants, surveys, interviews and usage data, in order to validate or invalidate many of the assumptions that had to be made on the impact of the design on its users when the operational version of EIES was originally designed. The unfortunate paradox that the evaluators have faced up until now is that the design group is always ahead of them in time. Our history since 1976 has been one in which design changes have preceded evaluation. In a real sense the design process has been the tail that wags the dog.

Fortunately this situation is now changing. One significant aspect of what the evaluation has been able to confirm is the diversity of human preferences both with respect to the interface and the styles of communications that take place. EIES began with a well-specified design of five different interface protocols and three well-specified alternative communication modes (messages, conferences and notebooks). However, implicit in the technological aspects of the design was the premise that individuals and groups would ultimately seek to tailor their communication and information processes around their tasks and objectives. To facilitate this capability, the design and implementation team has developed a rather unique computer programming language (INTERACT) to fully control the communications between any individual or group and the basic EIES system.

This capability allows the user to create or tailor his or her own interface; to superimpose data structures on free text; to create specialized delivery, scheduling or retrieval of communications on a selective or conditional basis; and to incorporate other computer aids as an integral and coherent part of the communication process. Since EIES represents a dedicated mini-computer environment, this ability is not meant to allow the incorporation into EIES of either very large structured data bases or number crunching computational capability. However, this ability would allow a group of designers and potential users of a data system to design and construct an "architectural" model of a data system, and to actually experiment with a sample of the data base, before a commitment is made to a paper plan for a large scale system. This is only possible because of the integration of the communications into the modeling and experimental process. For the evaluators the opportunity now presents itself to conduct more tightly-controlled experiments into specific current and potential aspects of interactive systems as well as computerized conferencing systems. Already two such experiments are under way: one dealing with human group problem solving and the other with memory retention and estimation of communications on the part of individuals. Recently a group of cerebral palsy-afflicted youngsters and residents of an old-age home have been added to the system to experiment with the ability to tailor the interface to a user population having rather severe limitations.

Uses/Users of EIES

In terms of scientific and technical groups, two of the new groups have begun to take advantage of this capability of selective tailoring and structuring of communication and information processes in rather major ways. The group of state legislative science advisors have their own select group menu which deals with the delivery and creation of technological and policy inquiries and responses. The very specific data format for these one-page items is superimposed on standard EIES text items and special indexes created to track what has been seen by whom and what is related to what. The starting computer-naive member of this group sees only the one menu that replicates the paper and mails system that this organization started with. He or she is free, however, to leave that interface, representing a closed subsystem within EIES, and move to the EIES standard interface when desired.

A group representing an industry panel responsible for setting industrial standards in the electronics area has created a subsystem called TERMS where members can propose alternative definitions for a single technical term. Individuals are notified of new terms and definitions and after reviewing they can vote on the definitions in order to gain quick insight into where they have a group consensus or disagreement, as a mechanism of focusing the discussion in their conferences. One specific panel is already dealing with eighty terms after a one-month effort. In contrast, the usual process of mails and bi-yearly meetings has required as much as two years of time to accomplish the development of an agreed-upon set of term definitions for a standards-setting committee.

One thing designers of interactive systems have felt in-
stinctively, and which appears to be emerging in a more definitive sense from the evaluation work and experience with
EIES, is that users span three overlapping classes: those who
desire the capability the interactive system offers to such
an extent that they will utilize any interface provided, no
matter how incompatible it is with their own cognitive process; those who have no motivation to use the system and who
will not, unless forced to by their job situation; and those
who actually can be influenced or swayed as to their degree
of use by the characteristics of the system and the design of
the interface. Much to our chagrin, the non-motivated class
appears to be larger among the user groups we have initially
been dealing with than any of us expected. It may be that
the reward and incentive systems in science are less attuned
to informal communications than they were in days past and
that the old images of the scientific process die hard. It
may very well be that those people who agreed to be part of
an EIES user group in the formulation of the original proposal and then never signed on discovered a significant difference in their images and their reality for scientific communications. This class of users has little utility for the
designers of the system at the technological level; however,
by their sizable population they have much to say for the
role and function of EIES-type systems within science. This
class of non-users will not utilize these systems until they
are provided some of the secondary scientific communication
processes such as review, publication and consulting. All of
these are of course technologically possible and have been
demonstrated from the point of view of feasibility. For
EIES-type systems to succeed with this class of users, the
informal communication process must become formalized within
the discipline--paradoxical as that may sound.

When it comes to our committed class of users in terms
of either individuals or groups, what we are beginning to see
is tailoring of the system in rather diverse ways as a function of interests and objectives. If we look at that group
as an empirical base it would appear that they lend validity
to the premise that what we are ultimately after is not one
best system in the optimum sense. Rather we appear to have
the phenomenon of local optima that are largely self-tailored.
While it is true that this small minority of the users are
often the only ones to utilize some of the more advanced
tools that EIES has to offer, this does not allow us to infer
that such tools are not of significant value because only a
small percentage of the population use them. A simple example
is the use of anonymity or pen names, which I believe must be
evaluated on other grounds than usage rates. In terms of

more complex examples, many tools developed by the more
advanced users become part of the EIES library so that they
represent added capabilities for other less-advanced users.
One such example is a survey design and execution system
which allows a user easily to format a questionnaire to be
sent to other users, with the automatic accumulation of the
responses in the author's notebook. The real challenge for
the design effort with respect to this group is to keep the
system at a level where these advanced users never reach the
saturation phase and where at least some of their accomplishments can be filtered down to those following in their foot-
steps.

It is really the class of users between the two extremes
that represent the greatest challenge for both the design and
the evaluation effort. EIES was not designed to provide a
single "average" system that represents some sort of com-
promise design to maximize this population of users. Rather
the intention is to provide an interface which allows them to
get started in some meaningful communication with others at a
minimal level of learning with respect to the mechanics.
However, it is also the intention to tease such users with
the knowledge of more power available if they should discover
the need to master it. In other words, our real objective is
to foster migration from casual use to the incorporation of
EIES as part of an individual's communication-information
processing behavior. I don't think we are very satisfied as
yet that we have been able to arrive at a proper balance in
the design between these two objectives. At the moment our
balance in design can perhaps be best described as a pre-
carious one that does not seem to work for all the users we
had hoped it would accommodate. This is an area where the
evaluation effort can provide considerable input to the design
process. It also probably offers the evaluators the biggest
challenge in separating out the design issues from the motivational ones. My impression is that our most irritated new
users are the computer sophisticates who are often taken
aback by the elementary nature of our introductory material
and the initial menu interface all starting users are provided. We have not at this point paid as much attention to
the computer sophisticate as we perhaps should have, in the
belief that they could more easily fend for themselves.

One significant new feature, which should improve our
ability to accommodate a wide range of users through the
initial phases of learning and facilitate migration to more
advanced knowledge of the system, is the ability to record a
user's input during an interaction and to send that record of
the input to others to play back as if it were their own

input, so they can learn by observing another user's interaction. This is about as close as one can come to sitting
next to another user at the terminal as a way of learning to
use advanced features.

Conclusion

There seem to be today two distinct paths in the area of
utilizing computers to facilitate human communications. One
concept seems to say that because communications is a generic
activity common to all humans there must in principle exist
one optimum way in which to employ the computer in this process. Given that one can evolve that optimum system, then
all users will use one system very much like we all have access to a telephone. I would tend to characterize most of
those working in what are termed "message systems" as having
this view implicitly if not explicitly. Certainly a company
developing a marketable system would like its potential
customers to believe that it had found the perfect system.
The foundation of our efforts with respect to EIES is very
different and somewhat common to those who characterize
their work as "computerized conferencing." It is our belief
that the introduction of the computer allows us to structure
and tailor the communication process to the objectives of the
individual and the group. We are not dealing with a more
efficient replacement for the letter, telephone or inter-
office memo but with a new communications medium that has its
own unique impacts on individuals and groups. EIES itself
has been designed very much as a laboratory without walls
where any geographically-dispersed group can reform EIES into
its own mold and attempt to evolve the design that seems to
satisfy its particular needs. While this is the belief we
have in the system, it will still be the job of the evaluators to tell us how well we reached this mark.

Part 3

Social Impacts and Policy Issues of Electronic Communications

13. Problems of Accountability in Large Federal Databanks

Two Views of Accountability

Holding record systems accountable has taken two different directions in the United States. (1). The most significant progress has been made at the individual level emphasizing the notion that record systems must be accountable to individuals about whom records are kept, and especially when those records are used to make important decisions about that individual. Enforcement of this principle of accountability to individuals is the keystone to several pieces of legislation such as the Privacy Act of 1974, the Fair Credit Reporting Act of 1970, the Fair Credit Billing Act, and the Family Educational Rights and Privacy Act of 1974. Enforcement of this principle of accountability to individuals is accomplished in these various pieces of legislation by mandating government agencies (and credit institutions) to acknowledge the existence of record systems (openness); to provide indi- vidual access to records upon request; to provide opportunity for review, for copying, for amendment and appeal of the contents of individual records; and to provide legal status to individual citizens in Federal courts to sue agencies for actual damages.

There is little doubt that the framers of this legisla- tion sought to maximize the fairness of government record keeping and decision making based on these records, and, given the immense size of these record systems, establishing by statute individual rights with respect to these records was thought to be the most efficient means of causing Federal agencies to re-examine their record policies, and if neces- sary, change them to improve record accuracy, timeliness, and completeness.

While it is perhaps too early in the history of this leg- islation to make definitive judgments, it is apparent to most that attempting to hold very large public sector record systems accountable by investing citizens with individual rights of access, review, appeal, and even legal status to sue, may not be the most efficient means to the goal of fairness, and is most certainly not the most effective. Predictably, Federal agencies have failed to keep reliable data on Privacy Act requests, but it is apparent that most requests originate from employees (about 95% of D.O.C. requests and Civil Service Commission requests). Given the four billion Government records in existence, there were around 15,000 requests three months after passage of the Act, but this has slowed to a trickle with only the Department of Justice experiencing a steady annual request for records (around 35,000 per year). (2). There have been a handful of appeals and requests for amendments to records, and even fewer successful court cases, largely because of the difficulty of proving damages.

The failure of the access provision to work has meant that Federal managers of large record-keeping systems have made few discernible changes in their day-to-day procedures for assuring accuracy, timeliness, and completeness, and in general, for assuring fairness to individual citizens. According to the Privacy Protection Study Commission report, agency managers claim they have always striven for accuracy, that completeness and relevance have different meanings in different contexts, and, interestingly enough, few if any agencies have even attempted to specify accuracy standards for their specific records. (3). And, as the President's Reorganization Project discovered, managers of Federal agencies have simply failed to take the Privacy Act seriously, they have not made it part of their operational decision making, and indeed see the Act as a hindrance to normal operations. (4).

The reasons commonly adduced for the lack of citizens exercising their new access rights are most interesting. The Privacy Protection Study Commission believed that citizens did not request access to their records for two reasons. First, the Commission felt that citizens cannot conveniently find out what record systems exist because the mandated, annual system notices are published in the Federal Register, and the annual compilation of these notices is about the size of several Manhattan telephone directories, in which few logical principles of categorization operate. Second, the Commission felt that the lack of access requests was a positive sign that Americans in general felt great confidence in the record keeping practices of the Federal government. (5).

These explanations of the failure of the single most important enforcement vehicle in the Privacy Act, namely individual access to records, only serve to obscure the fundamental

weakness of seeking accountability of huge data systems
through enhancing individual access rights. As it turns out,
and as decades of experience in other fields has taught us,
providing citizens rights without similarly providing the resources necessary to exercise those rights is no remedy to a
problem at all. In the area of government records, specifically, the legal minds who framed the privacy legislation seem
to have ignored the mundane realities of social life in which
only a few citizens have the resources of spare time, knowledge, writing skills, and cash to pursue individual access
rights in a manner sufficient to hold large Federal record
systems accountable, let alone to ensure their fairness.

Management Responsibility

This leads to consideration of the second principle
means by which we, in the United States, attempt to hold record systems accountable, namely through attempts to collec-
tively define limits on record systems operation, which, if
they are exceeded, could subject the agency or the system
managers to legal sanction, and which constitute benchmarks or
criteria by which Congressional oversight committees and the
public as a whole can judge the operation of a system. For
instance, the Privacy Act of 1974, as well as other more specific privacy legislation, mandates limits on the collection
of information by Federal agencies, the use of information,
the disclosure of information to other agencies, and attempts
to spell out the specific responsibilities of managers with
respect to information practices. Critics of the Privacy Act
point to these attempts at limitation as providing little more
than weakly-worded license to Federal agencies to go on doing
about the same as they have in the past. (6). Limiting agencies to collect only that information "necessary to accomplish
a purpose of the agency required by Statute," and limiting use
and disclosure to "routine uses" as defined by past agency
practice, hardly seem like limitations at all. The Privacy
Protection Study Commission itself concluded that "so long as
society countenances a particular line of inquiry, questions
as to how far it may properly go seem largely aesthetic." (7).
Briefly, it appears that efforts to hold Federal information
systems accountable by placing stringent limits on collection,
use and disclosure are doomed by the society's even more pas-
sionate desire to obtain efficient administration of Federal
programs. (The exception here appears to be information regarding individual exercise of First Amendment rights which,
by consensus of all parties, ought to be beyond the purview of
Federal data systems.) What this means is that as long as we
believe in efficient Federal tax administration, few interests
in society are willing to place stringent limits on internal
revenue agents' access, or use of information about indivi-

duals which may be helpful to efficient tax administration (once again, the exception here is First Amendment activities.)

Thus, efforts at holding record systems accountable by imposing limits on collection and use are vitiated by the lack of agreed upon meaningful limits. The same may not be true in the area of management responsibility. Here, criteria of judging the performance of records systems may b.: attainable and we turn now to consider the difficulties of obtaining accountability in this area. (8).

Management Responsibility:
System Quality

The omnibus Privacy Act, as well as other specific privacy legislation, impose upon management certain responsibilities in the area of system quality. Attempts to hold systems accountable, with respect to system quality, has taken two approaches. The first is concerned with the security and confidentiality of records and record systems. Under existing law, managers of Federal record systems are required to insure security and confidentiality of systems and to protect them against threats and hazards which could result in harm to individual citizens on whom information is kept (Section 3(e) (10) Privacy Act of 1974.)

The second approach to system quality is concerned directly with record quality: managers are required to maintain records which are accurate, timely, and complete to assure fairness to individuals about whom a determination may be made (Section 3(e)(5) Privacy Act of 1974).

There is little doubt that most attention in the area of system quality has focused on security and confidentiality. The successful physical and computer program penetration of social security record systems by the GAO, newspaper articles concerned with the diversion of criminal justice information to organized crime along with other articles on successful "computer crimes" in financial institutions, and so forth, have helped to create a thriving cottage industry for management consultants and computer scientists concerned with encryption, threat and risk analysis of various groups from terrorists to wayward insiders, to foreign powers. I was once asked, for instance, to write an analysis of what would happen if the Russians came into possession of a large Federal databank following a successful invasion. About the best I could come up with was that most of the individuals in the record system would be terminally inactive.

Without wishing to diminish the problem of security and confidentiality, recognizing that very significant system leaks are occurring due to insider activities, nevertheless concern with security and confidentiality has unfortunately diverted attention from a more intractable, costly, and significant problem for the average citizen, namely record accuracy, timeliness, and completeness. The more significant and frequent threat to citizens appears to be the routine creation, use and dissemination of poor quality records, and not the occasional failures of security.

Examples of Record Quality Problems and Methodological Difficulties

This section reports the results of investigations of record quality in three systems: the Supplemental Security Income Program (SSI) of the Social Security Administration, a large State criminal history file (which will be a major participant in the proposed FBI Interstate Criminal History Program), and the Internal Revenue Service file on individual tax filers. Also included here are references (but no tables) to studies of error in private sector billing systems. The major thrust of this section is to establish that the problem of record quality is significant and much larger than is commonly assumed in the scientific and lay communities. Second, the methodologies for studying this problem are not well defined.

Example I: Supplemental Security Income Program (SSI)

The SSI program was begun in the early 1970's as a Federal supplement to State programs of assistance to the aged, blind and disabled. The SSI program reflects the full complexity of large-scale Federal databanks: the program is funded by the Federal government, but it is administered by the autonomous State agencies who make determinations about eligibility; eligibility for the program's benefits requires extensive documentation and proof from medical records to earnings history; the records themselves are among the more complex in the Federal civilian sector because they depend upon multiple third party submissions (from doctors to hospitals to employers), and the records must be maintained for very long periods of time, essentially the life of the individual.

In 1977 there were slightly over 4 million beneficiaries who received a little over 6 billion dollars in assistance. In large part, because so much money is involved in this system, we know more about data quality than in the other systems. The Social Security Administration conducts monthly surveys (n=4500) of the beneficiaries in a quality assurance program. Early studies, up to about 1975, found errors in the program running about 20% of the case load file which included overpayments, ineligibles on the roles, and underpayments.

Tables 1-3 present the results of the survey period: April-September 1977.

Table 1 suggests the probability of a beneficiary case experiencing an error is around 13.4% nationally, and ranges as high as 17% in Boston and Denver (the error rate 95% confidence intervals extend to as high as 20% in these cities.)

Table 2 finds that around one-half of the case errors results from beneficiary submission errors, and the other half from agency errors either in data quality or decision quality.

Table 3 further breaks down payment errors (data on case errors is not available).

Because of the interest in detecting and preventing fraud in beneficiary programs, we know more about certain kinds of errors in these systems than in other systems. However, there are several methodological difficulties involved in interpreting these studies from the point of view of accountability, and Privacy Act provisions for accuracy in records.

The SSI data is based on random samples of beneficiary cases and beneficiary payments. This method seriously under-

Table 1. Case Error Rates with Confidence Limits (%)
April-September 1977

Region/Nation	Case Error Rate	Confidence Limits	
Boston	17.1	15.5	- 18.7
New York	12.5	10.9	- 14.2
Philadelphia	15.1	13.4	- 16.8
Atlanta	13.8	12.6	- 15.0
Chicago	11.0	9.6	- 12.3
Dallas	11.5	10.0	- 13.1
Kansas City	12.6	10.1	- 15.1
Denver	17.3	13.8	- 20.7
San Francisco	14.8	13.1	- 16.5
Seattle	10.1	8.5	- 11.8
Nation	13.4	13.0	- 13.9

Source: Systems Report (01-26-78)

Table 2. SSI Case Error Rate by Region – April–September 1977

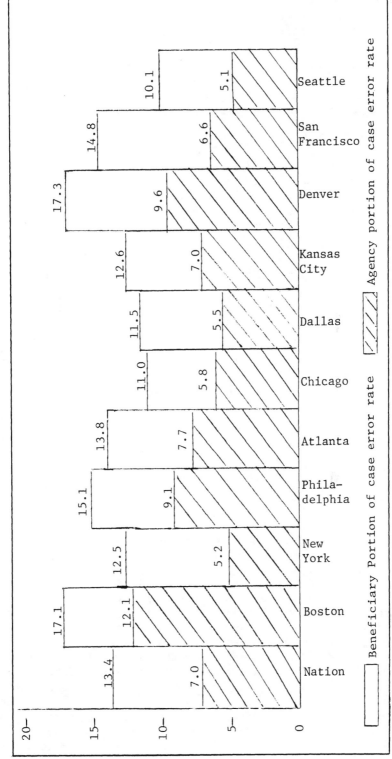

Source: CRE.45, CRE.83, CRE.84 (01-20-78)

Table 3.   Sources of SSI Deficiencies -- National

	Payment Error Rate % (10/76-3/77)
I. Agency	
Incomplete development, verification	2.5
Incorrect determination	.4
Incorrect data transfer to system	.2
Failure to follow up on change	.1
No DO action on Reported change, PE change	.1
II. Program Administration	
Policy interpretation variances	.2
III. System	
Systems processing	.4
Conversion	.05
Systems failure; interface other Federal programs	.2
IV. Beneficiary	
Inaccurate/incomplete information	2.0
Failed to report change in circumstances	2.3

estimates overall error rates because it does not include eligible persons who were erroneously denied benefits, does not include cases selected for review of auditors, active cases in the appellate process, or cases under investigation for fraud or other violations. To get an idea of the size of this problem, consider that about 340,000 cases were involved in the appellate process during 1977, around 10% of the beneficiary group. And no one knows how many beneficiaries perceived errors but did not appeal. Second, these investigations do not adequately distinguish between errors in underlying records and record processing, from decision errors of employees. This problem will arise in other examples. (For further documentation, see "Quality Assurance Report, April-September 1977, SSI Program," Social Security Administration, December 1978; and "The Year in Review," annual report of the SSA, 1977).

Example II: A State
Criminal History System

The second set of data reported here is based upon a
study of record quality in a large State criminal history system (New York). The importance of this data lies in the fact
that central computerized State repositories for criminal
records will be the building blocks for a Federal interstate
message switching system proposed by the FBI, and which is
currently undergoing Congressional scrutiny. (9). Under current plans, the FBI would establish a message switching capability to enable states to query criminal records in other
states, and hence the quality of records in state systems will
determine the quality of the proposed national system.

The data reported here are drawn from research conducted
by the Legal Aid Society of New York and the sociologist
Richard Faust (see Tatum v. Rogers. S.D.N.Y. 75 Civ. 2782).
They are based on a representative sample of 793 criminal his-
tory records, involving 2741 arrest events drawn from the New
York State Criminal Justice Information System.

Table 4 presents some of the overall findings on record
quality. As indicated, about 25% of the records were complete
and accurate; 45% of the records showed arrests but no formal
disposition (charges dismissed or dropped, acquitted, or found
guilty). Nearly 10% of the records were just plain inaccurate
when compared to local court records of known accuracy; another 10% were incomplete; and another 10% were ambiguous.
The findings that only 27% of the records in a large state
criminal history system are complete and accurate may seem
outlandish but it is in correspondence with the results of investigations of other State systems, as well as FBI studies of
records submitted by the states. (10). The findings also seem
outlandish in light of the fact that by 1985 nearly one-half
of all criminal justice decisions made in the United States,
from arrest to parole, will be assisted by the proposed national criminal history system. (11).

Even with random samples of records and extensive costly
validation procedures, difficulties emerge in the interpretation of these data. Clearly, not all record errors lead to
decision errors. Some record errors may be of little consequence to the determination about an individual, and some record errors may be corrected by lower level personnel who in-
formally adjust their decision making. Generally, however,
random samples of record error probably underestimate decision-making errors. First, a single record error will likely
lead to multiple, erroneous decision errors. Thus, the wrong
person may be arrested, and then follows a series of linked

Table 4. Accuracy and Completeness of Disposition Information (for 2,741 prior arrests since January 1, 1973)

Disposition Checked in Court Records, Rap Sheet is:	81%	
Complete and Correct	27%[1]	
Blank	45	
Incomplete[2]	8	
Inaccurate[3]	9	
Ambiguous, showing "Combined Charges"[4]	10	
Case Pending: No Disposition to Report	9	
Court Records Sealed or Otherwise Unavailable	3	
No Docket Number Locatable[5]	7	
Total Percent	100%	100%
Total Number of Prior Arrests	(2,741)	(2,210)

[1]The sampling error for a figure of 27% in a sub-sample of
2,210 arrests is about 2%, i.e., the percent "complete and
accurate" is between 25% and 29% (see Table 1).

[2]Incomplete entries lack full information on convictions: (1)
the sheet indicates conviction of an "Attempt" but the specific crime is not stated; (2) the sheet indicates a sentence
but does not list the conviction charge.

[3]Inaccurate entries were of three types: (1) basic disposition
was incorrect (conviction listed for dismissal, or vice
versa); (2) multiple listings of a plea to more than one
charge when there was a plea to a single charge; and (3) sentence was recorded incorrectly.

[4]Ambiguous entries show several charges in the disposition
column with indication of plea to the "combined charges," when
the plea was to a single charge.

[5]There are two reasons why a docket number would not be locat-
able: (1) the charges were dismissed before arraignment; (2)
the docket number was listed in the court records under a
different name from the one in the defendants's rap sheet.

Source: Affidavit of Richard Faust. Tatum v. Rogers,
S.D.N.Y. 75 Civ. 2782 1978.

decisions by the prosecutor, probation, judge, and correction
personnel. Second, incomplete but correct data will lead to
additional errors. For instance, police officers may arrest
and/or detain a person on the basis of three prior reported
arrests even though no formal conviction ever resulted, in the
belief that "someone with three arrests must have done something wrong." Here, correct but incomplete records will lead
to organizational decision errors. Third, programming errors
and bugs will inevitably lead to additional errors in decision making. Briefly, not much is known about the man-machine
interface in such systems, to determine with certainty the relation between record errors and decision quality.

Example III: IRS Tax File Errors

The final table to consider is concerned with the nature
and frequency of known data quality problems in the Internal
Revenue Service's individual tax return file. The IRS file
results from over 90 million taxpayer submissions in each
year; this file is retained for five years, and becomes the
basis for IRS decisions concerning audit, investigation, and
final tax determination. In this sense the IRS is different
from the FBI file in that it is truly participatory: 90 million citizens are allowed each year to more or less tell the
truth about their earnings by creating their own files. Yet
the IRS file is not unique; the Social Security Administration
files are similar in their reliance upon data submitted by
citizens, employers, and other third parties.

Errors in participatory file systems like the IRS are
difficult to analyze because citizens routinely introduce errors into the data base, sometimes by intention and often by
making accidental mistakes. Additional difficulty is presented by the IRS itself which has not been forthcoming about
its record quality problems. To get some idea of detected and
estimated errors, consider the published data in Table 5: of
82 million annual returns filed in 1976, 1.6 million (2%) were
audited (all per annum), 7 million (5%) were unpostable because of irregularities, 8.2 million (10%) contained mathematical errors. The IRS estimates there are an additional 4.8
million (6%) who failed to file or did so erroneously. Altogether, about 27% of the records are thought by IRS to have
problems and no one knows, after all the correction procedures
have been taken, just what correspondence there is between IRS
records and reality. (12).

Perhaps the most interesting point is that despite these
difficulties, IRS management must make a tax determination,
and hence, is forced to treat the records as if they are correct. Briefly, dealing with large record systems and basing

Table 5. Known Record Quality Problems in IRS Tax File

of 82 million records filed in 1976:

Irregular	5%
Problem Accounts	10%
Math Errors	10%
Audited Accounts	2%
Estimated Erroneous and/or non compliant	6%
Total:	33%

Source: "Summary of Major TAS Features – Present and Proposed Security and Privacy Measures," Internal Revenue Service, 1976.

decisions upon them requires a certain suspension of critical judgment. Rarely, but in interesting ways, the suspension of critical judgment and resulting decision errors surface through citizen protests in the courts. In Neal v. U.S. for instance, the IRS was unable to give an account of why its computer had misappropriated a 1973 tax refund and applied it to a 1971 account, or why its system refused to accept the protests of the taxpayer but instead doggedly harassed him (Neal v. U.S. 402 F.Supp.678 (1975)). And in another case, a court-ordered examination of the IRS records found around 60% of the IRS-labelled non-filers, who receive special investigative attention, were in fact deceased, married, or unemployed (U. S. v. Liebert, 383 F.Supp.1060 (1974)).

The suspension of critical judgment, not recognizing underlying data quality problems, not recognizing the incomprehensibility of much of their own software, leads inexorably to acceptance of the system reality as the only reality. (13).

Example IV: Private Sector Billing Systems

It would be unfair to conclude that only public sector systems are difficult to hold accountable because of system error: private systems are experiencing the same difficulty as are private citizens involved in these systems. In a forthcoming article to appear in Communications Magazine, a publication of the Association for Computing Machinery, Professor of Computer Science Theodore Sterling reports the results of a sample of consumer households in Western Canada. The findings of this survey suggest that the probability of consumers

experiencing an error in automated billing systems is around
.4, that of experiencing exactly one error, around .25 to .33.
Credit card, utility, department store, and mail order bills
are the most frequent institutional source of error. Billing
errors are the most common type of error, followed by bank
transactions, and in one-third of the documented cases, four
or more separate contacts are required to correct errors.
These findings are in rough correspondence with previous published surveys and numerous, proprietary, unpublished surveys.
(14).

There are several methodological difficulties involved in
interpreting the results of this and other similar surveys.
Reliance on self-reporting of errors is fraught with problems
of recall, recognition of error, and self-report bias. Perhaps a more difficult problem is that such surveys tend to
confuse system decision quality with record quality. System
decision errors may occur for reasons having nothing to do
with record quality, but instead with a variety of man-machine
interface points within an organization, or with client failure to comply correctly to billing notices. Caution is required when interpreting the results of such surveys. Nevertheless such data are useful in gauging perceived frequency and
nature of system decision errors, and may point to underlying
data quality problems.

Cultural Bias Against Investigation
of Record Quality Problems

Computer systems, like other institutions, do not routinely record their mistakes. Managers of some large Federal
systems, moreover, are resistant to inquiries about record
quality. When approached on the subject, they are either shy
or defensive, and as it turns out, no Federal databank now
operating has conducted a systematic investigation of record
quality. This situation is changing very rapidly, however, as
Congress refuses to fund major system developments proposed by
Executive agencies which cannot give an account of record
quality, and hence, which cannot be held accountable.

Unfortunately, the bias against study of this problem ex-
tends into the scientific community as well. A few years ago,
with a colleague of mine in computer science, I submitted a
proposal to a large Foundation which sought to investigate
record quality problems in a number of small public and private systems, ranging from university student record systems
to local criminal justice systems. The small grant was turned
down on the basis of two outside reviews. One, from a social
scientist, was revealing.

This critic argued the following:

-computers are neutral instruments which do not, of themselves, magnify the indignities which an organization can cause citizens to suffer.

-the proposed examination of record quality in computer record systems is in itself indicative of conceptual bias.

-computer record systems are more accurate than manual systems, and one should do a comparative study of manual and computer record systems, not just a study of computer systems.

The ideas that computers are neutral instruments, and that manual systems have more errors than computer systems, are themselves interesting hypotheses which may or may not be true (depending on the quality of record conversion, computer files often simply automate existing manual file errors. This appears to have happened with criminal history records). Even if true, they are somewhat irrelevant and largely of academic interest. To be sure, the credit reckoning of the local grocer and the manual files of the local sheriff may have contained many errors. But unlike national credit and criminal history systems, the local manual files of years ago had less important rewards and punishments to allocate, and errors were, in any event, localized. Also, errors which did occur were most likely easier to detect and correct than those occurring in large national systems. Moreover, the question is not whether we will have manual systems as opposed to computer systems, for clearly we will have large national computer systems.

The most important questions then are what is the frequency, nature, and origin of error in large computer record systems; what management policies can reduce error and at what cost; and what new national information policies can we create to attain acceptable levels of error (recognizing that all systems will contain some irreducible level of error).

Policy Implications: Are Record Quality Standards Possible?

The experience of DHEW in requiring State Welfare files to attain specific levels of accuracy as a requirement of Federal funding indicates that stricter standards of management responsibility for record quality can be achieved and that establishment of record quality standards for specific systems is a feasible route to greater accountability of systems to both the public and Congress. Yet, managers of Federal systems are probably correct when they assert that record quality standards must be adjusted to the circumstances of each operating system. Our exploratory research indicates that some systems, because of great expense and/or excellent management practice, are relatively error free, while others not similarly situtated are error prone. It appears that systems operating within a single organization, in a competitive environment, dealing with low complexity transactions, and where service failure is highly visible, are likely to have lower levels of error than systems with opposite characteristics. The extent to which management procedures can overcome inherent environmental limitations (like record complexity) and at what cost, is largely unknown.

While our knowledge in this area is virtually non-existent, several interim policy measures are feasible as a means of strengthening accountability of large Federal systems which play an important role in the life of millions of Americans:

  1. Creation of internal, record quality audit groups.

  2. Public reports on a routine basis indicating the distribution of record quality problems and management efforts to solve these problems.

  3. Training of lower level organization personnel in record quality detection methods and formal notification of employees regarding detected record problems.

The alternative to these interim measures is continued and even increasing reliance upon systems whose social and legal consequences are unpredictable.

Notes and References

1. For another view of accountability in large record systems, see Kling, Rob. Five models for the social accountability of computing: a brief review (mimeo). Irvine, CA: Department of Information and Computer Science, University of California; 1978 January.

2. Privacy Protection Study Commission. Appendix IV: The privacy act of 1974: an assessment. Washington, DC: U.S. Government Printing Office; 1977.

3. Ibid: 53.

4. Federal data processing reorganization study: operational management team report (draft). Executive Office of the President; 1978 June 15.

5. Privacy Protection Study Commission. Appendix IV: 33.

6. Mowshowitz, Abbe. The conquest of will. Reading, Mass.: Addison-Wesley; 1976.

7. Privacy Commission. Appendix IV: 16.

8. For additional discussion of management responsibilities and ethics in the operation of large record systems, see Sterling, T.; Laudon, K. Humanizing information systems. Datamation; 1976 December.

9. Sterling, T. System errors and their impact on end users (forthcoming). Communications of the ACM. 1979.
Laudon, K.; Sterling, T. Frequency, origin and consequences of error in large systems (forthcoming).
Laudon, K. Complexity in large federal databanks. Society/ Transaction; 1979 May.

10. A national survey of the public's attitudes towards computers. Montvale, N.J.: American Federation of Information Processing Societies; 1971.

11. A preliminary assessment of the NCIC computerized criminal history system. Office of Technology Assessment, U.S. Congress; 1978 March.

12. LEAA, a principal funder of State Criminal History Systems, has never conducted a data quality study of these systems. Partial examinations will be found in the following references:
Criminal record systems. General Accounting Office; 1973 January. (most examined records in California found

inaccurate or incomplete).
New York Times; 1974 October 27.
Proposed criminal history program. Federal Bureau of
Investigation; 1976. (45% of records found incomplete in
survey of 800,000 criminal histories).
The American criminal history record. Sacramento, California: Project Search, Inc.; 1976. (criminal justice
personnel perceive 31% of records incomplete, 10% erroneous.

13. Costs and benefits of the comprehensive data system program. Washington, D.C.: Institute for Law and Social Research; 1975.

14. Internal Revenue Service. Summary of major TAS features -
present and proposed security and privacy measures. Preliminary analysis of the IRS tax administration system.
Washington, D.C.: Office of Technology Assessment; 1977
March.

15. The difficulties of comprehending very large programs is
discussed in Weizenbaum, Joseph. Computer power and human
reason. San Francisco: Freman; 1976.
The extent to which lower-level programmers can make sub-
tle changes in programs unknown to the original planners
is discussed briefly in Kling, Rob. Electronic fund transfer systems and quality of life. Montvale, N.J.: AFIPS
Conference Proceedings, National Computer Conference,
Vol. 47; 1978.

14. Social Change Through Electronic Communications

Introduction

Depending on how you date it, we're either 25 or 30 years deep into the computer revolution. Certainly, in terms of the way in which large organizations have adopted and used computers, it's been since about the mid-1950's that computers and their associated communications systems have begun to move so heavily into our organizational and daily lives that hardly anything happens, if you just take a moment to think about it, where there isn't some force of computers and telecommunications driving it. Our credit standing is checked electronically, there are all the kinds of applications in insurance and employment and health care, taxation, motor vehicle registration and licensing, police service and so forth; it's a fair thing to say that our lives are all pervaded by the force of electronic communication and electronic calculation. One could also assume that by now we have not only a pretty good technology but also a pretty good set of techniques for assessing the impact of this powerful new set of tools on society. I hope I'll persuade some of you by the end of the time I finish speaking that we really don't have such tools today, and that we're struggling very hard, 30 years deep into this process of change in society, to understand how to assess technology; to understand the problems of definition, the problems of measurement, and the problems of values that are involved in it.

One has to start with some historical sense of the way in which we have tried to acccomplish this assessment over 30 years, what some of the pitfalls have been. Perhaps we can see some of the directions which history has opened up for us today, that we can at least see as alternatives or perhaps even as some vision of the Holy Grail. Perhaps we can see some congruence of ideas at the end of this.

Technology Assessment Problems

Let me suggest what the main problems of technology assessment in this area are, as I see them, before turning to
some specific efforts to describe how assessments like this
have been done. First of all, you have the problem of defining and establishing what the empirical baseline is, what
the state of nature is, if you want to think about it that
way. Computer systems enter the worlds of organizations,
whether it's a bank or an insurance company, a police department, or the Social Security Administration. The first thing
one has to do is have an accurate, empirically accurate, snapshot of how that organization functioned before the utilization or the adoption of various kinds of computer techniques
and telecommunication options. That is by no means an easy
matter, because social science is a poorly developed tool in
our country and throughout the world. One doesn't exactly
reach on the shelf and get a kind of description of how the
FBI really works, or how the Social Security Administration
really works, or how IBM or Bank of America really work. If
it was as simple as that, if we could simply go as scientists
can - look at the combination of several chemical elements
before the introduction of a new element - that's good science. We are able to come up with a satisfactory description
of chemical interactions before a new agent is introduced, but
it is difficult in the world of social reality, with organizations and governmental processes.

There is just the raw problem of information. The way
things really work in the private domain is often proprietary,
or is kept from our full view, because these are institutions
that can claim for themselves under our law various rights of
confidentiality and privacy. They don't have to conduct themselves under fishbowl openness, and they don't. If you really
want to know how X, Y or Z Corporation is dealing with marketing, is deciding upon strategies of sales, and the way in
which decisions are really made inside the organization, in
the private sector, it is very difficult to come up with that
kind of baseline information. Even in the public sector,
though we start off with the idea that what Government does is
the public's business; even though we have many techniques
such as the competition between the legislative and executive
branch to ventilate, supposedly, what the administrative arms
of our Government are doing; and even though clearly, in comparison with the private sector, we do know a great deal more
about what the FTC or SEC or OTA is doing, the fact is it's
still very difficult to get the kind of quality information
about that kind of operation that would satisfy anybody that
an adequate baseline, an adequate snapshot, of the pre-comput-
er, pre-information technology introduction had been well
done.

I don't mean it is impossible, but just think for a minute about the FBI, for example. Anybody who wrote histories or studies of the FBI 20 years ago could only hint darkly at the fact that maybe there were certain things that were going on that we did not know about. Happily, these people continued to write books so they are not embarrassed by the fact that new things come out, because they write about them as soon as they do. But if you took a look at literature on the FBI in the 1950's or early 1960's, clearly it didn't begin to have in it some of the things about the techniques the FBI was using, some of the goals that the FBI moved into in the early 60's, middle 60's, late 60's or early 70's.

It is a very difficult problem to establish that kind of baseline so we can begin to identify the impact of information technology in the organizational world. Well, how do we do it? Essentially, it has become a rather artistic endeavor. One has to take any particular function in society such as health care, or any particular single organization such as a corporation or the Social Security Administration, and in effect work up a very careful baseline description of what information was collected by that organization for carrying out its functions: service functions, management functions and so forth. One has to look at the technology by which that information was manipulated, a combination of eyeballing, accounting machinery, and early ADP equipment. One has to see the flows and movement of the information in the organization. One has to distinguish between real signals and noise: what information is used by people to make real decisions as opposed to what is just an endless collection of information for the purpose of filling up file drawers.

The last thing, of course, one can do is ask anybody, and expect to get a truly useful answer, "What information is really important to you and what information do you need the most, and what information matters for your decision-making function?" I remember in the middle of the late 60's there used to be a kind of standard scenario in which bright-eyed systems analysts went to the counties and cities and many other organizations that were taking on third-generation computer systems, and said: "And tell me, sir, what information matters most to you in your job here at the old ABC Agency?" They expected to get a useful answer, or honest answer, and then to come up with redesigns of information flow. First of all, many people couldn't have given an honest answer even if they wanted to; and why they would tell a stranger, the kind of consultant who borrows your watch to tell you what time it is, any kind of honest answer about the power uses of information inside an organization is ludicrous. Yet that passed for some system design information-building in some phases of the use of computer technology in organizations.

I think the way one has to try to do this is to create
the kind of sensitive, very carefully- and painfully-construc-
ted schema of what the information function is in any given
organization that you're studying; or what it is across organizations, if you can take a look at the whole line of functions. Creating that kind of baseline - just to show you another problem that is involved in that - is not even something
you can do once because, while you say I'll take as my first
snapshot a look at the baseline of how information was used in
that organization before the first adoption of computer technology, you then have to deal with the fact that one doesn't
adopt computer technology once and for all - it isn't a kind
of one-time process. There are new generations of computer
systems, changes in the configurations, in the way in which
hardware and software and people are connected to one another.
You have to go through a series of continuing snapshots of the
changes in the relationship of the new collection and usage of
information technology with the changing organization itself.

So, the first kind of problem that I think one has to
identify and struggle with is the development of an empirical-
ly-oriented descriptive picture of the organization or the
function whose use of information technology one is trying to
assess. It's a difficult process, but I don't think it is
impossible. It really calls for the bringing together of insight into the rules and realities, the political realities as
well as the formal realities, of the systems with which organizations use information.

The second major problem that we have great trouble with
is trying to control all the other forces that are at work
while we are trying over a period of time, often 5, 10, 20
years or so, to assess the impact of information technology in
the organizational world. That is, we live in a period of
social turbulence, a time of dramatic and rapid economic
change, political change, social change, value change in our
society. At what point we say an organization has done this
or has not done this because of technology, as opposed to
changing social notions of what is legitimate for organizations to do or not to do, becomes a very difficult matter.
Even calculating something that would seem to be as relatively
easy as whether an organization's power has been enhanced for
good or for ill by the introduction of computer technology becomes very difficult.

I'll give you an example. I wonder, not in the sense of
a public opinion poll, but in the sense of rather informed expert judgment, whether one should rate the FBI today as more
or less powerful than it was 10 or 20 years ago. At first
blush, one could take the pure technological answer. They

have powerful computers, they have elaborate telecommunications systems. You can now punch in a license plate or a name into a terminal in Sacramento, California, and get back in a matter of seconds a flash whether someone is wanted or not, or whether a bond has been stolen or a car has been reported stolen. The arresting officer can have information that used to take 24 hours or 48 hours to be obtained. Yet I would think that it is a rather questionable matter whether the FBI is more powerful today, because of the weakened prestige of the FBI compared to what it was under J. Edgar Hoover's way, and given the lack of willing cooperation of large segments of the population, black and white, radical and reactionary, with certain kinds of Government agencies and functions. For example, I think the informer network is much weaker today than it was 20 years ago for the FBI, and on the whole, if you deal closely with criminal justice functions, you learn that one informer is worth two computers in terms of the real way in which law enforcement ever finds out anything in a society. So that being the case, it's a rather tricky measure to try and understand how much power has been enhanced by the introduction of large computer systems into organizations. It may or may not be true that power vis-a-vis the individual, power vis-a-vis other institutions in the society, power vis-a-vis intermediate groups in society, has been increased or decreased, when you correct for other social, economic, political, cultural changes in society that are taking place across the period in which you're trying to do this kind of assessment.

Third, I think you have the problem of the value framework for judging what kinds of changes have taken place and whether you think they're bad or good. That is, I think it's the beginning of just honesty, much less sophistication, to say that technology assessment is not a value-free enterprise; we're not here measuring chemical reactions, we're measuring changes in a social system and whether we think they are good or bad depends on our perspective. Therefore, when you say that the computer has caused this change or that change, almost inevitably the very way you say it carries with it the notion you think the change is good or the change is bad. That gets back to which values in society you hold, who you think deserve to be praised and who condemned, and has to do with all the allocations of privilege and power and caste in the society. Even in the very nature of what elements you choose to make part of an assessment, you are advancing your own social view of the society. In other words, I don't think it's quite true that you can create a value-free assessment of technological impact. You can hope to make it open and honest, and you can provide a description of why you think a change has been of a certain character. But life has changed. No organization holds still over a decade or two

decades while we try to look at the impact of the computer
systems used in that organization, or in that society. So,
whether you think the impact has been good or bad depends on
how you relate to the changes that have taken place in that
society, in that organization, in the delivery of health care,
in the tax function, or whatever, in the society, over that
time. I think we need a very open and honest way of calculating how we are dealing with the value framework in the selection of terms and the ways in which we described changes that
have taken place.

An Application of Technology Assessment

There was once an interview with George Bundy when he was
in the White House as an aide in the Kennedy Administration,
after the first 75 or 80 days of the Kennedy presidency. It
was a time of great exhilaration and media love affair with
Camelot. I remember an interviewer asked Bundy, "And how do
you think the Administration has been doing so far?", and
Bundy said, "Well, I think we're a little like the Harlem
Globetrotters in the performance they put on before the game;
they pass the ball under their legs and under their arms and
they throw it down the field - it's all very dazzling - but,
at some point the whistle blows, the game begins and then what
matters is whether you make baskets."

So far, I've tried to give some sense of what the problems are. Let me try also to suggest that, at the end, it's
the matter of making baskets. Can we do this? How do we do
it? I tried to do it a couple of years ago. What I would
like to do is run through very quickly the framework that I
found useful for trying to apply an analysis of the overall
effects of the use of information technology on democratic
institutions. That was, in itself, as I say very openly, a
value-charged way of setting my terms for assessment. That
is, I presupposed that what I was interested in was the operation of democratic institutions. That carried with it the
assumption that I was talking about constitutionalism, about
due process, about responses and responsible government, about
distribution of power. When I tried this, I spelled out very
carefully what I thought were the publicly-accepted, honored
concepts of democratic institutions, always subject to the
realities of the way things work as opposed to the way we
would like them to work. Assessing against that concept of
democratic institutions, I tried to come up with a number of
points at which I thought we'd had enough time 25 years into
the computer revolution to try to make some large judgments
about this. The particular terms that I used were these: I
selected four dimensions of democratic institutions.

The first was efficient government. Here I was talking about things such as the planning and evaluation function in organizations, their decision-making processes, the ways in which oversight functions were handled inside organizations, and the way in which various kinds of administrative-efficiency judgments about centralization and decentralization could be dealt with.

Secondly, I looked at the areas of individual rights, and in particular, I was concerned with privacy and confidentiality, which I link together as being the aspects of what information is collected about people for making various kinds of social judgments about them, and once collected how well an organization can keep its promises of confidentiality. That is, is it able to use the information in ways that were agreed upon in the individual bargain and which respect the conditions under which people have either willingly or under requirement of law given information to various kinds of organizations? The other aspect of individual rights was the due-process dimension which includes both the ability of people to know what's in their records and to have mechanisms for questioning or correcting or protesting against what is there. Also to the extent that the use of information in organizations is seen to follow the kind of known pattern that gives people in the classic, due-process sense the feeling that the procedure by which a decision is made is just, it has elements of equity and fairness about it, and that people are treated in equitable ways. That, I think, is at the heart of what we hope to accomplish with due-process mechanisms in our constitutional system.

The third dimension of this kind of assessment was the distribution and balance of power in the society. What I was interested in looking at here was how much the use of computers seems to have changed and in what ways it had changed the constellation of power relationships between public and private sectors within the various units of government - legislative, executive, and judicial. To the extent that certain assumptions about distribution of power were ones which were thought to be enduring ones in the American scene as opposed to things that were easily changed, depending on the kind of economic moment we were at, or the kind of way in which our population was made up, I tried to relate these to major continuing assumptions about the American political system that might or might not be affected.

The fourth dimension - it seemed to me a valuable one - was citizen participation. Here I was interested in looking at the way in which the access to and freedom of information in the society had fared under information-technology development, and, as another aspect to that, the way in which the

exercise of participative rights seems to have been affected
or not affected by information technology. The reason I put
it that way - the exercise of participative rights - is that
I've never shared the enthusiasm of those electronic afficion-
ados who think that somehow if we all have a button in our
living room and we can be asked at 8:00 in the evening, "Shall
we make war on Cambodia?", or "Should the President raise farm
crop support?", that somehow that represents the participation
that a literate society wants to pursue, even though the technology might become available. There is another aspect, too,
which is the right not to participate - which happens to be a
civil liberty in our society that we all ought to cherish very
carefully. You can get a 99% vote turned out if what you make
clear is that if people don't vote, they may get shot. We can
have "ja" and "nein" plebiscites, but I think our assumption
has been that the right to vote "no" by not going to the poll
is still part of the democratic society. We may deplore its
lack of responsibility and insipid participation, but I vote
for a society in which people are not going to be dragooned to
the polls if, in fact, for all kinds of reasons, they don't
want to go. So I think the exercise of participative rights
or options is the way I would describe that element of the
assessment.

Results of the Assessment

In the interest of time, let me just try to give you some
very broad judgments that I reached about these four sets of
dimension as to how information technology has fared, or how
we as a society have fared with it, so far. Since it would
take a lot of time to give a lot of examples, and a lot of
qualifications, I don't want to apologize for not giving them,
but I'll just try to suggest that they are there and you can
ask me about them later. We start first with efficient government. On the whole, it's my sense from having been involved in looking at a lot of Federal agencies and organizations in varying ways that the major impact of the computer
and of communications systems so far has been essentially on
service delivery, on the large high-volume types of information movements in the tax system, the motor vehicle system,
licensing systems, and so forth, where armies of clerks were
necessary to shuffle pieces of paper repetitively so that
various kinds of calculations could result in a social security check getting out, or so that someone can now have a computer program to see whether, if you are an accountant and you
claim these deductions on your income tax, you should have
your form kicked out for an audit or not. So the correlation
of various kinds of statutes with information can be very
nicely and usefully accomplished by computer technology.

I wish I could say I think a lot more has had real impact in the organizational world than automation of high-volume repetitive functions, but I'm seriously doubtful of it. Let me go through the other places where one hears that there has been a lot of impact and I will indicate why I feel fairly negative about them. In terms of the planning function, management information system functions: there are very few organizations that I know of in which computers are more than simply an aid to the smarts of top management, whether it be business management, or university management, or Federal agency, civilian or military management. This is all you can say that the modelling, simulation and management planning function amounts to. What I mean by that is, not that there isn't very sophisticated and interesting computerized modelling and simulating systems; it's that, when I've gone into organizations and studied how key decisions were actually made, and looked at the context of the decisions, two things usually happen. First of all, the actors will say, "At the certain key point we all agreed that the model could not take those certain critical pieces of information into account, so we corrected for that in the way in which we made our decision." The model was simply not long enough, it was not all-encompassing enough, or simply that the information was almost of a personality kind that one could not quantify and that a key decision of a bank, a key decision of a planning agency, could not really be coped with by the model for the kind of executive decision that was necessary. As a result, the model was very useful and effective in confirming the broad trends that the management needed to know and wanted to know, but if you get to the heart of the decisions that make the difference as to whether a business is successful or whether an agency pursues a program effectively, and so forth, I haven't seen where models and simulation computing have really had the kind of decisive effect that, in theory, they're supposed to have.

The second reason is that I've noticed, inside executive suites both in the public and private sectors, that modelling becomes a part of the political competition of executives. That is, a model doesn't exist sort of in the abstract. It is part of the power struggle over whose philosophy of the business or the agency or whose direction of it is going to prevail. There can be debates over the meaning of the model, the meaning of the output, the meaning of the simulation which returns to the political dimension, the political domain, the decision-making process. In a number of places where I have had a chance to look, I could just see that what really had happened was that two conflicting factions of the organization had each come up with an interpretation of a strategy based on what the model had said and the real decision had been one, as always, in which one set of policy forces had prevailed for better or for worse, and that the model would seem to be simply another piece of evidence, another piece of data, that could be weighed ultimately in the political sense. The decision was going to be made in the political sense.

So I think that we have not, inside government organizations or inside private organizations that I am familiar with, had a significant transforming effect on the executive function. You might say, "Well, that's to *be* expected. We're dealing in the Himalaya Heights of decision making, maybe it's very complicated and tricky and political and so forth," but how about further down? There was a great deal of hope that computer systems would make program evaluation much better, that one would be able through computer analysis to come up with an awareness much better than before of whether our welfare systems were working well or not, or our health care systems, or all the kinds of examples we could think up. Again, in the interest of time, let me just say that I'm extraordinarily unpersuaded that we're doing a better job of program evaluation inside agencies as a result of computerized resources. Not that we don't have lots more data - of course we do. But program evaluation, it seems to me, is again fundamentally understood from a political perspective. The major questions are: Who is to be served by programs; at what costs, compared to who else is now being well served; where the competition takes place; and what kinds of information are involved? If it's too uncomfortable to live with, the main function of an organization is to see to it that that information is not produced, or if it is produced, not in such time that it will hurt the agency with legislative oversight or the next election.

I once talked to a very shrewd city manager in California who said, "There are two clocks that run in our community; one is realtime and that's what's on the wall, the other is political time." Political time is the period within which we can live with information if it becomes public, and our job is to see that information doesn't get out in political time before we can cope with it. Anything we have to do by way of managing information, to see to it that we are not embarrassed in that way, is really what a lot of our whole activity goes into. Let's take health care, for example. I think that for all the advantages of computer technology, we have not been able to make the kinds of decisions in health care that go to producing a better way of treating the individual patient in a health care institution with information about that patient. That's not to say that we don't have in certain situations excellent capacities to do certain kinds of diagnostic things with data bases or that there have not been important other studies done, but if you look at the problem of the person in the health care system as a whole, we still have a labor-intensive health care system. The computer has been excellent

in lab test automation and in certain kinds of billing and
accounting procedures, although still most hospitals will tell
you they're behind in the capacity to bill the Government for
what they are owed or to control their costs more effectively.
For all the bold talk about hospital information systems and
patient information systems, the computer has simply not had
an effect upon either the evaluation of programs or the creation of new program mechanisms inside health care.

So on the whole, without taking more time, I would go
through one by one the key functions inside the definition of
what is efficient government or efficient organizational behavior, and I would say that on the whole the computer has
been a superb faster typewriter, a faster accounting machine,
a faster disseminator of checks and bills and statements, and
it has been obviously very important in terms of identifying
certain trends that then can be pointed out to middle management and top management for certain kinds of organizational
responses. But much beyond that, I remain a fairly profound
skeptic about the impact of automated information systems on
the other functions of the management art. I think that as
long as we are at this stage in the computer system revolution
in which organizations, for the most part, retain their iden-
tities - for example, in government I think what's much more
important today is that the computer has not changed Federal,
state and local competitions and rivalries - it has not really
yet revolutionized the functional divisions of the way law enforcement is conducted or taxation is conducted. It has not
overcome the good and the bad elements of the distribution of
function and the separation of function in American governmental systems. This is the remarkable negative statement to be
made about this aspect of the impact of the technology.

Let me turn secondly to individual rights. As far as the
impact of the computer on individual rights, looking first at
privacy and confidentiality, the dramatic point is that computer technology arrived not in a placid decade, a time when,
say as in the 1890's or 1920's, American values and trust and
confidence in large institutions would have made the computer
be viewed probably as a kind of unalloyed blessing. Rather,
the vital point to note is that computer technology spread
through the world of organizations at a time when we were just
beginning to undergo profound changes as to who was really entitled to the benefits of rewards and opportunities and privi-
leges. In the 60's and 70's, there were profound changes in
attitudes about religion, race, sex, and sexual preference,
dress, and so forth, and the trouble was that all of the record systems were full of the old disqualifications and the old
notions about using information to reject people for all kinds
of benefits and opportunities. I think that in one critical
sense, the struggle of the last 20 years has been whether the

computer would be used to reinforce the old morality, the old judgmental system of a society, or whether there would be some way to make computer systems also able to reflect the new changes on political rights, social rights, and racial rights.

Much of the struggle over the way in which computers are being used or should be used has been between, really, three camps in the U.S. First were those that liked the way things were being done before; who thought that, on the whole, those people who deserved to be on top were on top and therefore they were running the society, and the computer was a way of keeping it running the way they wanted. Secondly, there were those, to give it a quick characterization, the kind of Fortune 500 enlightened leadership in the business community, types who said, "No, the subjects will get much too restless and the machine would get kicked too much and the plug might even be taken out, if we don't find a way to show people that the use of information can be responsive to new equal employment opportunity concepts or political rights concepts." They represent what I would call the system adjustment approach to the privacy and confidentiality issues. This is the fair-information-practices approach which is, of course, that we can see that information practices are fair with computer technology or anything else. The thing to do is to define properly what information goes in and give people a right to see what's in their record.

There is a third approach which I think is the one that is at war today with the system adjustment approach, which is a social change position. This means that it is not just satisfactory to have fair information; one should be active in trying to discipline the use of information in order to achieve a more egalitarian society, a more civil libertarian society. That is why I think the value perspective comes into this kind of assessment of impact, because I think that in the Privacy Act of 1974 one could see both strains at work. Senator Ervin and the Senate Subcommittee on Constitutional Rights clearly had an anti-surveillance, pro-civil liberties, First Amendment thrust to their definition of what should go into the Privacy Act of 1974; whereas Elliott Richardson, the ex-HEW Secretary and eloquent witness before the committee at that time, was, I think, representing through the fair-information-practices approach the notion that what we really wanted to do was to see to it that people could be at peace with the machinery, would not feel abused but would feel that there were mechanisms by which fairness would be provided. I think that we are still very much in that struggle and this is really what the impact of the computer on information privacy amounts to. It involves the question of whether the organizations that now have the capacity to use computer and communications systems for handling information are going to be

allowed to do it as long as they live up to certain kinds of
informational fairness practices, or the extent to which there
would be some points at which organizations may be told by law
or as a matter of social pressure that they cannot build certain systems at all, or that they can't link certain systems
together at all, or that people are not going to want a certain function done more efficiently because of "privacy" considerations.

I guess I might have started off this discussion about
privacy by saying that I don't believe that the use of the
computer automatically transforms the privacy relationships of
organizations and individuals. It doesn't inevitably mean
that because you get a computer, an organization will automatically collect more personalized information about people or
exchange it more widely than it did before, or that inside the
black box will be things that we will never know. My assumption at the beginning of my comments is that, if we want to,
we can have a variety of controls over the use of computer
systems in the privacy sense, but that the choices of ours in
society do not flow automatically from the adoption or not
adoption of computers. Some of the most pernicious information systems I know are entirely manual; there are secret manual files, there are dirty manual files, and there are nasty
manual files. It really doesn't have as much to do with the
medium as much as it does the organizational intent and use.
Therefore, I don't start off with the idea that computer systems represent some kind of Attila the Hun function imposed on
the democratic society. The individual steps up to a problem,
i.e., a computer problem.

I think to provide what we mean by the traditional due-
process notion it takes a different set of mechanisms than
simply saying that if there's an error in your bill or there
is some problem in your rap sheet we will provide human beings
to work it out with you, and we'll show you, for example in
an amended printout, that it's correct. For most people,
keeping track of where the information is stored about them is
not really a traditional activity, and an intervening offer to
correct the information about themselves is not a normal activity.

Let me try to put it in a kind of class-term sense for a
moment. Concerning the well-educated in our system, the people who have complicated tax returns done by their accoun-
tants, I don't really tremble for their fate under the computer information society, but I think protection of them is important too. The people I'm really worried about are the less
well-educated and those in the society for whom being able to
make these corrections when information is used about them is
not normally within their competence. They don't have the

kind of individual resources, or, I am sorry to say, the organizational resources often enough, for doing this. That is,
in the late 60's and early 70's, there was a good deal more
organizational resource available to the poor and to racial
minorities and to various other groups than I think we have at
the moment. I think it may be an agenda problem for the future.

Somewhere, the money and the organizational dynamics may
be needed to give individuals who, because of lack of various
kinds of language resources, income resources and class-based
disabilities, need to have advocacy and intervention on their
behalf where information is used in ways that can affect their
interests. I think we lack some of these countervailing organizational resources by which to navigate the impact of information systems in governmental and business life.

On the whole, then, looking at individual rights, I think
that where we are today is that the computer systems have not
as yet transformed the fundamental relationships of the individual and organizations in the privacy and due-process sense;
that we are still able to set, either singly or in combina-
tions, a fair-information-practices, due-process, constitu-
tional-rights approach to these kinds of controls over information systems. On the whole, I suppose I'm fairly optimistic
that for lots of reasons, the privacy and due-process issues
have been very well publicized, that organizations believe, I
think accurately, that the American public could cause them a
lot of grief if they don't pay attention to it. I think
Lou Harris, for example, now gets quite a lot of his income
each year in being able to tell business people at his private
conferences for them that, in fact, the American public is
very worried about privacy and doesn't feel happy about computers, and that various big insurance companies or big banks
ought to pay a lot of attention to this. Whenever a problem
gets to that kind of point, there is some hope that there can
be some effective legislative, judicial and other remedies.

We shift to the third area that I was talking about, distribution and balance of power. Here, I think I start with
the assumption that computers and communication technologies
are very expensive affairs. To have a computer and the space
for it and the technicians to operate it is not something that
the local Black Panther's organization or the local ACLU Chap-
ter or other kinds of non-corporate and non-governmental institutions are easily able to harness for themselves. In the
very basic sense, I think one starts with the understanding
that computer systems have been so expensive and so technologically innovative that they have in their first several decades been primarily the playthings of the well-organized and

the well-financed in our society. For quite some time to
come, the expenses of the large computer systems make this
something that will continue. That means, then, that the existing distribution of power according to wealth and organizational resources was reinforced, it seems to me, by the arri-
val of computer systems. Certainly, if you take a stark contrast, the Molotov cocktail is a power-redistributing technology as opposed to a power-enhancing technology; in that sense
computers are a power-enhancing technology.

On the other hand, things are happening that could make
it a little different in terms of distribution of power. For
instance, it is really astounding when you think about the
hand-held calculator and the capacity to move that much computing into the hands of large, large numbers of people very
cheaply. I don't know what better social decisions we make
from those hand-held calculators, but at least as an example
of how you can bring computing power to a very different kind
of scale of cost and accessibility, it certainly is a dramatic
example. Let me use, as a perhaps quick single example of
what I think our choices are here, an example that Ralph Nader
has presented several times. If you wanted to use computer
technology for the consumer of the United States, why not do
it this way? There should be terminals available every couple
of blocks. A person should be able to walk in, and if he
wants to buy a rug or eat at a restaurant he should be able to
punch in the name of the local carpet merchant or the local
restaurant and find out whether the carpet store has been
cited for violation by the Better Business Bureau or by the
Federal Trade Commission or any other of the oversight agencies that protect us, or whether the restaurant has just been
put on the suspended list or probation list for having too
many roaches in its kitchen. If you wanted to have information at the hands of individuals for making their choices effective, then why don't we use computing power that way?

Ralph Nader commented that he was not holding his breath
until that kind of information system was put in for two reasons. One, it would be a system that in some way would have
to be voted by the taxpayer through the Congressional process
or the counterpart at the local government level. He was not
convinced that the alliance between business and government
would permit that amount of useful information being made that
immediately available to the average citizen. Second, he
thought that in many ways the American consumer is so trans-
fixed by the consumer-group society that, although some people
like to read the consumer reports and other kinds of publications that in effect try to help on a much less efficient
scale, that the political mobilization of the American consumer needed to make that kind of information system possible

was something that he was skeptical about. Not that he didn't
try, but he was just very worried about the capacity to build
the political support so that that kind of a system would take
place.

In short, what I'm suggesting is that, when you look at
the choices that are made as to who builds a computer and for
what purpose, we have really two mechanisms: one, the market
system, which means where large entrepreneurs see a buyer's
market, a profit market to do it; and secondly, in the governmental system, it's a matter of political organization and
political skill. On the whole, as a kind of bottom-line
statement, I think that information technology has been a
power-enhancing or a power-consolidating part of our political
system for the first several decades. This is pretty much
short of larger changes in society than I see at the moment,
and where it's likely to be for some time to come.

Finally, let me say a word about citizen participation.
I sort of anticipated this in the beginning. On the whole, I
think information technology has had very little effect on the
quality or the nature of citizen participation in our society.
First of all, as far as freedom of information is concerned, I
did a study a few years ago in which I looked at about 40 or
50 Federal agencies to see whether the computer usage that
they had done had affected their ability to comply with the
Federal Freedom of Information Act. I not only asked the people in the agencies themselves, where one could get certain
kinds of predictable answers, "Yes, we can still comply with
the law just as well with our computer systems as before," I
also asked the information seekers, various kinds of minority
groups, civil liberties groups, reporters and the press, Congressional committee staffs and a wide variety of those who
want to get information from the Federal government. I think
that what was very clear on the basis of the study that I did,
and I think it would be just as true today as a few years ago,
is that in terms of freedom of information, what really matters is the law defining what the public has a right to see or
not and the interpretations of this given by judges and so
forth, not the use of computers and the capacity to store the
information electronically. There are a few examples that
were given where the absence of software to draw out the information that a given group wanted has posed a problem, but
even then that's been solved to some extent by the 1974 Freedom of Information Act amendment requiring indexes and other
kinds of support mechanisms by which people can know what is
in computer systems. I don't think that, so far, information
technology has made it either much harder or much easier for
the press and the public and the interest groups that serve
the public to get from governmental institutions the infor-

mation that is needed for making decisions. While I could
imagine a number of ways in which we could be better at that
than we are, I don't think there has been a major negative
effect by the use of computer technology.

On the other aspect of this, the last one that I put in
that category, which is the opportunities of citizens to engage in participation, the kind of electronic townhall, the
electronic plebiscite concepts, have been not only unrealized
but deservingly so. That is, I think democracy is a very complicated system. It calls for a variety of overlapping mechanisms of group education and staggered political decision
making and party mechanisms and interests groups, without trying to suggest that it's a wonderful Rube Goldberg work that
was sort of struck by the brow of God, or something like that.
Making it electronically direct is not going to enhance the
quality of that democratic process. While we might talk about
lots of other things that might make American society more
democratic, I don't see in the near future that if, somehow,
we all get home box offices and two-way cable and we can be
asked at 8:00 what our view is on something and punch it in,
that that will be a major useful change in the quality of
American participation. Inside organizations I think there
are many interesting aspects of participation that could be
enhanced by electronic means, but let me not go into that because I really want to stay on the question of the electoral
process, the political process. Here, I think one has to distinguish between delivering more information and getting better citizen participation. If I thought it was information
delivery that we were talking about, I could be optimistic,
but I don't think it's that. I think it has to do with recognition of interest, articulation of interest, and the mechanisms by which people are brought into the political system.

Let me close then with a short comment. Recognizing all
the problems of what we want to include as the standards for
making assessments of the impact of computers and communications systems on democratic institutions, my overall judgment
would be that a lot less has happened than some of the people
who look at these relationships might think. Some of what has
happened I suppose I would view as being valuable. Some of it
is perilous, but we are early in the art of really understanding what we will eventually decide was the impact of information technology on the larger set of social and cultural and
political forces in our society. I think all of us who work
in the field can use all the help we can get in both termin-
ology and techniques in trying to draw up the balance sheet.

15. Policy Options for the Future: Congress, the FCC and the President

Introduction

For most persons, the term telecommunications means little more than plain old telephone service; and a telecommunications policy issue is whether the telephone company should be allowed to implement one of its recurring increases in basic telephone rates.

Telephone service is indeed the largest and most universal telecommunications service, directly involving nearly every person on a daily basis. Moreover, U.S. telephone service is among the best in the world. We can take considerable pride in both the quality and universality of this service, and in the telephone industry that brought this about over the past 100 years. We can also take pride in the record of scientific and engineering accomplishments of the Bell Telephone Laboratories, one of the finest research and development institutions in the world. And we can further be thankful for the fact that the U.S. telecommunications network, unlike that of most developed nations, has been a beneficiary rather than victim of the major armed conflicts of the past 50-75 years.

But telecommunications does not end with good telephone service, and even the best present-day telephone service is itself capable of much improvement and enhancement. Thanks to work done both by Bell Labs and others, there is a large and growing number of telecommunications and information-processing equipments and services beyond those normally associated with plain old telephone service itself, as specialized devices and services of particular applicability and benefit to specific business, government, and other institutional needs. Of these, quite a few - such as telephone-answering machines, solid-state computer devices, electronic games, and so on - have already become public consumer items, and many more will soon follow. Still others, such as airline reservations

systems, electronic cash registers, cashless transaction systems, and the like, have become of considerable indirect benefit to the general consumer, and these applications are also
increasing at an ever-growing rate. The current telecommunications policy debate concerns in particular these non-conven-
tional equipments and services, although it surely is heavily
influenced by concerns for the continued availability of high-
quality, low-cost telephone services.

Industry Legislative Campaign

About three years ago, the Bell System led the telephone
industry in launching a massive, grass-roots campaign whose
purpose was to convince the public and the Congress that telephone companies alone, operating within exclusive franchise
areas free from competitive suppliers, should be authorized to
provide all forms of telecommunications equipment and services
for public use. The industry contended that this was Congress' intent when it enacted the Communications Act of 1934,
and that unless prompt legislative action were taken to "re-
affirm" that intent the quality and availability of basic
telephone service in the U.S. would rapidly decline while its
costs to many citizens would sharply increase. All these dire
consequences would result, claimed the industry, from growing
competition in the telecommunications business as a result of
decisions by the Federal Communications Commission (FCC) and
the Courts over the past 20 years.

This campaign, launched at the peak of an election year,
initially attracted a significant number of Congressional en-
dorsements. It also elicited strong rebuttals from the FCC
and others, who pointed out that previous industry forecasts
of such dire consequences had proven unfounded; that the industry was currently enjoying record revenues and earnings;
that the current industry forecasts of adverse consequences
resulted from the pancaking of a series of theoretical, worst-
case assumptions that had an infinitesimal probability of occurring in practice; and that these forecasts were further and
fatally flawed both as to methodology and input data. As a
result, the industry's campaign did not gather sufficient mo-
mentum to secure passage of the proposed legistation - entitled the Consumer Communications Reform Act of 1976 - in 1976,
and by the following year most observers agreed that these
proposals as originally presented were dead.

This campaign did, however, trigger what may have been a
long overdue review of this nation's statutory communications
policies and guidelines, under which these services have been
supplied and regulated since 1934. It is that review, which
has been undertaken by both the House and Senate subcommittees

on communications, that now serves as the forum for a significant debate over future government policy and regulation in
this field. Initial reactions by both these subcommittees indicate they are not persuaded that this increasingly vital
sector of the nation's economy and "nervous system" should be
entrusted to a monopolistic industry structure, and are in
fact likely to endorse and even expand the pro-competitive,
deregulatory actions of the FCC and the Courts. There remain,
however, many interested parties to be heard from, as well as
the difficult and complex task of converting whatever policy
principles the Congress endorses into specific statutory di-
rectives and regulatory mechanisms capable of ensuring their
effective implementation. To understand and appreciate the
significance of the options available to the Congress, one
really needs to comprehend the telecommunications and related
developments of the past 20 years or more that triggered this
review, as well as some of the interests and forces associated
with these developments.

Technology History

Stated most succinctly, this period has been one of increasing confrontation between dynamic technological developments and consumer needs, on the one hand, and a considerably
less dynamic but essential and well-entrenched telephone industry, on the other. The technological push has resulted
from a virtual revolution in electronics technology that began with the Second World War and has continued through the
Cold War, the Space Race, the Computer Revolution, and on up
to the present. The demand pull has been created by the increasingly complex communications and information processing
requirements of modern society, which have been fueled by many
of these same technological developments.

At one stage in its history, following the long-delayed
expiration of Alexander Graham Bell's original patents, the
telephone industry was itself both dynamic and competitive.
Most of the basic features of modern telephony were in fact
developed during this competitive era - many, such as central
office switching, by individuals and firms outside the Bell
System. By the 1950's however, the industry had convinced
both itself and much of the public, thanks in large part to
the efforts of AT&T, that life could be both more orderly and
more rewarding if each local telephone company operated within
an exclusive geographic area free from competition, and relied
entirely on a single long-distance network operated by the
Bell System for long-distance telephone services. For the so-
called "independent" or non-Bell companies, this arrangement
was sweetened considerably by the receipt of a disproportion-
ate share of long-distance revenues as compensation for the
use of local facilities to orginate and complete such calls.

Under AT&T's leadership, the telephone industry by the
1950's had also established effective control over the manufacture, ownership, and use of all equipment associated with
the telecommunications network, both through so-called verti-
cal integration of the manufacturing and service components of
the industry and through tariff prohibitions against any and
all forms of independently-manufactured, customer-provided
equipment. This effort was so successful that at one stage
telephone subscribers were even denied the right of placing a
plastic cover on their telephone directory, on the grounds
this was a violation of company property rights.

Another achievement in the telephone industry's efforts
to control telecommunications development and eliminate competition during the first half of this century was the creation of Federal and State regulatory agencies. While the pop-
ular conception is that such agencies were created over the
opposition of the industry primarily for the purpose of pro-
tecting the public, in the telecommunications field AT&T is
credited with recognizing that public regulation might be much
easier to live with and profit by than would competition.
Thus, both Federal and State regulatory programs were established with AT&T's active support, culminating with the Communications Act of 1934. This Act established the Federal
Communications Commission, set the scope of its jurisdiction
vis-a-vis that of the States, and established the basic public policy objectives for U.S. telecommunications development.

Neither the Communications Act of 1934 nor any other public action authorized or endorsed the establishment of a telephone industry monopoly in telecommunications. In fact, some
provisions of the 1934 Act clearly contemplated continued competition where it then existed, while other provisions charged
the FCC with enforcing certain pro-competitive features of the
antitrust statutes. However, the Act contained numerous loop-
holes and due-process features that could be, and were, used
both to suppress competition and to exploit monopoly positions. These included the ability of established carriers to
implement new services and rates without prior regulatory authorization through the nere filing of tariffs; to engage
prospective new carriers in prolonged regulatory and legal
proceedings whenever they attempted to enter the market; to
refuse interconnection with new carriers pending the conclusion of regulatory proceedings which the same established carriers could also prolong; and many similar defects. As a consequence of these and other weaknesses in both the theory and
practice of regulation, regulatory control of telephone industry rates and practices, including anti-competitive practices, has been far more myth than reality, both at the Federal and State level; thus the telephone industry, led by the

Bell System, was for many years able to establish, maintain, and reap the benefits from a de facto telecommunications monopoly.

Early Regulatory Decisions

The technological revolution that began with the Second World War was not confined either to conventional forms of telecommunications, such as analog voice telephony, or slow speed digital telegraphy, or even to the telecommunications field at all. These new technologies – such as microwave radio, solid-state digital processing, communications satellites, and so on – created opportunities for entirely new communications services, for significant improvements and enhancements of conventional services, and for new information-processing services which themselves created a demand for innovative new telecommunications equipment and services. In fact, they formed the nucleus of a major new industry built around information processing, as distinguished from information transfer, or "telecommunications." Many of these technological developments took place outside the established telephone industry, in aerospace and electronics firms engaged in space and defense research and development. Even where the Bell Laboratories pioneered in new developments, these did not always find ready acceptance or application within the traditional, voice-oriented telephone industry.

By the early 1950's, these various technological developments coupled with the general increase in U.S. industrial activity, social mobility, etc., had created a classic opportunity for major advances in telecommunications – a booming demand for new services, and an array of potential suppliers possessing the technical and managerial expertise and resources to satisfy this demand. Both these pressures, of consumer demand and of potential supply, inevitably focused on the monolithic telephone industry and on the regulatory mechanisms that had been created to deal with this industry when both needs and technologies were much simpler. Under these pressures, cracks began to appear as early as 1948, when the FCC decided that persons other than telephone carriers should be authorized to use radio systems to provide mobile radio-telephone services for hire. But it can hardly be said that the FCC rushed to authorize competitive sources of telecommunications equipment and services: in 1956, the Commission upheld AT&T tariffs which barred consumers from attaching non-electronic, cup-like devices called Hush-A-Phones, manufactured by an independent supplier, to their telephone instruments. Only after a Federal court held that such restrictions violated the public's right to use their telephone service in

ways that were privately beneficial and not publicly detrimental, did the FCC reverse itself and rule these tariff restrictions were unlawful.

Following the Hush-A-Phone decision, the FCC began to be
somewhat more responsive both to consumer needs and to potential new suppliers of telecommunications and related equipment
and services. In 1959, the Commission decided that private
parties having special requirements for long-distance communications, such as railroads, pipeline companies, and similar
organizations, should be allowed to establish microwave radio
systems for this purpose. This decision was made over strong
objections by AT&T and the telephone industry, who contended
that all such frequencies should be reserved for exclusive use
by that industry, on the grounds that this was necessary to
achieve both spectrum and economic efficiency.

In 1968, twelve years after the Hush-A-Phone decision,
the FCC was again asked to rule on the lawfulness of an AT&T
tariff prohibition against customer-provided equipment. The
particular device in question was the Carterfone, an independently-manufactured item which provided an acoustic coupling
between a consumer's private mobile radio system and his telephone handset. The purpose of the Carterfone was to enable
businesses operating fleets of radio-dispatched vehicles to
relay telephone calls to the operators of those vehicles,
without in any way affecting telephone network operations as
such. After concluding that the Carterfone in no way
threatened any harm to the telephone network or its subscribers, the FCC declared Bell's tariff prohibitions unlawful. To avoid further case-by-case litigation concerning each
new item of customer-provided equipment, the Commission ruled
that blanket prohibitions against such equipment constituted
an unlawful abrogation of the consumer's right to use his
telephone service in ways that were privately beneficial but
not publicly detrimental - the principle enunciated by the
Court in Hush-A-Phone. AT&T, and all its connecting carriers,
were ordered to eliminate such blanket prohibitions from their
tariffs. AT&T responded by filing new tariffs which replaced
blanket prohibitions with a blanket requirement that all customer-provided equipment be connected with the telephone network through carrier-supplied "protective" circuitry - at
additional charges to the consumer. Only after further, pro-
tracted regulatory proceedings and legal challenges were
these blanket discriminations against customer-provided equipment and independent equipment suppliers finally eliminated,
when the Supreme Court in late 1977 refused to entertain the
telephone industry's final attempt to justify its longstanding
practices. Some 21 years after Hush-A-Phone, consumers and

independent equipment suppliers finally achieved the rights to
which that decision held they were entitled, not just as a
matter of communications law but as a matter of Constitutional
principle.

Following the 1968 Carterfone decision, the FCC began to
pay increased attention to the concerns of both consumers and
prospective new suppliers of telecommunications and related
electronic equipment and services. A major concern was that
the Bell System, despite its leadership role in the telephone
industry, was not responding adequately to the burgeoning
needs and technological opportunities for specialized telecommunications equipment and intercity services needed to satisfy more sophisticated data communications and other requirements of modern business, government and social institutions.
Another concern was that, in attempting to expand into the
data processing business itself, the telephone industry might
unfairly destroy the viability of this highly competitive industry, while simultaneously burdening its captive telephone
subscribers through cross-subsidization of its competitive
activities from its monopoly service revenues. In a series of
decisions between 1969 and 1971, the Commission authorized new
classes of intercity carriers to offer specialized intercity
services via microwave radio and communications satellite
facilities, and established rules prohibiting telephone carriers from providing data-processing services except through
separate, armslength subsidiaries.

During my own tenure at the FCC, from 1972 through 1978,
the roles of competition, consumer choice, and consumer rights
in the telecommunications field were further expanded and de-
lineated. Among the major achievements during this period
were the previously-mentioned successful implementation of the
Hush-A-Phone and Carterfone decisions; establishment of the
telephone carriers' obligation to interconnect with intercity
systems and services provided both by consumers and by independent specialized carriers; elimination of Bell System tariff
prohibitions against the resale or sharing of telecommunications services; and establishment of a standardized methodology for allocating costs and setting rates among competitive
and non-competitive services so as to avoid unlawful cross-
subsidization. Each of these and many related rulings were
bitterly contested by the telephone industry both in Commission proceedings and in legal challenges. In virtually every
instance, the Courts upheld the FCC on both policy and legal
grounds. Moreover, despite repeated telephone industry forecasts of dire consequences for themselves and for the residential telephone subscriber from each and every Commission
action, the industry has enjoyed some of its most prosperous

years, while both the variety and use of telecommunications
equipment and services have expanded significantly with no adverse effect on basic rates.

Having been unsuccessful in its attempts before the FCC
and the Courts to halt the introduction of competitive forces
and consumer rights into what had become a de facto monopoly,
the telephone industry in 1976 launched the legislative campaign referred to previously. According to the industry, it
required non-competitive conditions to, among other reasons,
ensure technical and operational integrity of the telecommunications network and preserve longstanding pricing practices
and revenue sharing arrangements which allegedly benefit residential and disadvantaged telecommunications consumers at the
expense of long distance and commercial consumers. As noted,
the FCC and others point out wholesale inaccuracies and in-
consistencies in the industry's forecasts, as well as its
claims as to both the existence and benefits of alleged historical cross-subsidies. The Commission noted, for example,
that the Bell System's own specialized intercity services
could hardly be contributing to the maintenance of low-cost
residential telephone service, since rate investigations going back to the mid-1950's - prior to intercity competition -
had shown that Bell revenues from these specialized services
did not fully cover their costs. The FCC also noted that
AT&T's cost accounting system was totally incapable of identifying costs by service category, and its so-called "value-
of-service" pricing scheme was so grossly structured that no
beneficial cross-subsidization of needy subscribers by afflu-
ent subscribers could possibly be established - even if such
subsidization were deemed socially desirable. Moreover, the
Commission found no evidence that either the consumer-equipment market or the intercity-services market exhibited any
economies of scale that would make a single source of supply
either desirable or inevitable. Finally, the FCC noted that
the pace of technological development and cost-reducing innovation had quickened perceptibly since the advent of competition, with obvious consumer benefits and no apparent carrier
harm.

Ultimately, telephone industry spokesmen acknowledged
their inability either to determine the relevant costs and
profitability of particular telecommunications services, or to
demonstrate any technical or economic harm as a consequence of
existing competition in the consumer-equipment and intercity-
service business. With this, their public lobbying effort in
support of a statutory monopoly abated, although attempts to
preserve as much as possible of their favored market position
have continued. Meanwhile, both the House and the Senate

subcommittees on communications have undertaken the task of
sorting out the various claims and counter-claims concerning
the future role and effect of competition, monopoly, and regulation, both in the telecommunications industry itself and in
related fields such as data processing.

It should perhaps be noted in passing that telephone industry revenues and profits have continued to increase at a
rate considerably higher than the inflation rate, while no
rate increases in basic telephone services have been traced to
any competitive effects. Indeed, it appears that competition
may have stimulated significant increases in the use of telephone services themselves, bringing in additional revenues and
earnings.

The Industry in the Information Age

The acquisition, exchange, and use of information have
already become critical functions within an increasingly complex world society. Some authors have compared current developments in the processing and use of information with the Industrial Revolution, labeling this the beginning of an "Information Age."

To achieve the maximum public benefit from each and every
sector of this overall revolution in information handling,
much creativity and innovation will be required. Monopolistic
industry structures are not noted for great creativity and innovation, nor has public utility regulation proven capable of
stimulating such creativity and innovation on the part of monopolistic suppliers. On the other hand, competitive free
enterprise seems to stimulate and reward creativity, innovation, and efficiency - and to penalize the converse - both
generally and with particular reference to communications/
electronics developments. This argues for public policies
which maximize the opportunities for effective marketplace
competition in each and every sector of information/telecom-
munications development, and minimizes the need for and role
of traditional public utility regulation in these fields.

The draft legislation prepared by the House subcommittee
on communications appears to follow this course - at least
this surely seems to be its intent. There are, however, some
potential pitfalls which any new statement of national telecommunications and information policy must avoid if these objectives are to be achieved. For example, it must be recognized that the telephone industry did not overnight establish
its overwhelming dominance of all facets of the telecommunications business, nor the Bell System its overwhelming dominance of the telephone industry. After more than 100 years,

the industry's roots are deeply embedded in every aspect of this nation's social and economic structure; and for at least the past 20 years, it has demonstrated an almost fanatic determination to maintain the status quo as regards telecommunications industry structure - meaning no lasting competition within the industry.

There is a cliche which holds that the past is prologue to the future. That expression seems particularly apt with respect to the telephone industry and consequently to future telecommunications issues and developments. Despite the recent FCC and Court actions noted, the telephone industry still, to an overwhelming extent, dominates every facet of this business, including the competitive markets for specialized equipment and services - and seems likely to do so for many years to come.

In each sector of the business, from equipment manufacture to local exchange operations to long-distance operations and services and more, Bell System operations and revenues completely overshadow not only those of its competitors but indeed those of the entire telecommunications industry. As the only truly nationwide, fully integrated supplier of every form of telephone and other telecommunications equipment and services, the Bell System is far greater than the sum of its respective parts in its impact on telecommunications development and regulation.

Information processing has developed as a highly competitive, unregulated industry, and that structure seems to have resulted in a high rate of technological and service innovation with significant benefits to the consumer. The major public interest threat in this area appears to be potential monopolization, and the most likely source of such monopolization appears to be the telephone industry, which has attempted to encompass more and more data processing services within its telecommunication service offerings where anticompetitive cross-subsidization is virtually impossible to detect or correct.

Public Policy Options

This paper has concentrated on the negative aspects of Bell's prior performance, because these aspects constitute the most significant public policy challenges. The Bell System has, of course, made significant contributions to the development and supply of telecommunications and related equipment and services in the past, and can be expected to remain the major source of such developments in the future. Recently, under the spur of competition, the System has become considerably more innovative and efficient than in previous years.

The challenge to public policy is to ensure that Bell System
and other telephone industry participation in future telecommunications and information-processing activities is accomplished in a manner which maximizes their opportunity and in-
centives for constructive contributions, while minimizing the
threat of anti-competitive practices that would cause a return to stagnant monopolization.

There are a number of policy options that one could consider in attempting to deal effectively with this situation,
of which four seem worthy of discussion. One could simply retain the status quo, i.e., the Communications Act of 1934, the
FCC and Court rulings of recent years, the existing Bell System structure, and continuing regulation as now practiced by
the FCC. In one sense, this would constitute a pro-competitive, deregulatory policy since the Commission has in recent
years focused increasingly on market structure issues, with
a strong bias toward competition in lieu of public utility
regulation. However, the likely result of efforts simply to
maintain the status quo is remonopolization of the industry.
The FCC's resources, jurisdiction, and processes have been
taxed to and beyond their limits merely in authorizing the entry of competitive suppliers and establishing cost-allocation,
pricing, and interconnection guidelines for avoiding anti-competitive practices by established telephone carriers. To en-
force those guidelines in such a manner as to ensure that full
and fair competitive conditions actually exist and continue to
exist would require far more resources and statutory power
than the Commission now possesses. For example, the FCC presently has no jurisdiction over operations and services that
are purely intrastate, nor over data-processing services as
such. In ruling that telephone carriers could not prohibit
the interconnection of customer-provided equipment to the
telephone network, the Commission extended its jurisdiction
over such matters, vis-a-vis State jurisdiction, to the limit
on the grounds that interstate telecommunications services
over which the FCC has jurisdiction were being affected by
such tariff provisions. In ruling that carriers must enter
the data-processing business only through separate subsidiaries, the Commission could only register its concern that
such activities might burden the telecommunications consumer
with unnecessary costs, rather than any concern for maintaining competitive conditions in the data-processing industry itself, over which the FCC has no jurisdiction.

In the future, many new equipment and service opportunities may involve only intrastate activities, over which the
FCC has no jurisdiction, or they will involve activities in
the data-processing field which are so integrated with telecommunications that it will be difficult to separate them.
What is data processing and what is telecommunications? It is

very unclear how far the Commission's present jurisdiction
will go in dealing with those types of issues. Most State
regulatory authorities have been far less demanding of telephone companies, and far less favorable to competitive conditions, than has the FCC. In fact, several State commissions
have continued their efforts to thwart both interconnect
equipment and intercity-services competition through their
certification procedures, through gerrymandered rate structures for local exchange network and equipment services, and
other means. With its limited jurisdiction, resources and enforcement powers, the FCC could not possibly monitor Bell System and other telephone industry activities on a broad enough
basis to prevent anti-competitive practices and consequences,
even with State regulatory support. Without such support, or
indeed in the face of State opposition, the task becomes even
more impossible. Thus the long term prospect of attempts to
maintain the status quo is in fact a return to telecommunications monopolization by the telephone industry. Given the
growing opportunities for merging telecommunications and information-processing services, this remonopolization would
likely encompass the latter as well.

A second policy option, that is reflected in the initial
House legislative proposal, would be simply to declare all
telecommunications and information services open to full and
fair competition, and remove much of the FCC regulatory jurisdiction and control mechanisms. Proponents of this approach
sometimes cite the recent deregulation of the airline industry
as a model that might be followed. At first blush, such pro-
posals for total deregulation are appealing. Unlike the airline industry, however, the essential preconditions for full
and fair, unregulated competition do not currently exist in
the telecommunications and information-processing field.
Those preconditions include the existence of a reasonable number of comparably-sized firms, none of whom has effective control of the market, essential resources, or essential access
points to the market. In the airlines field, there were
several comparably-sized firms none of whom exercised or could
be expected to exercise such control. In the telecommunications field, the presence of the Bell System violates every
precondition for full and fair competition. Once again, the
likely long term result of throwing this field open at this
point to unrestricted, unregulated competition would be repeated and expanded monopolization of the entire field by the
Bell System.

A third policy option, which seems to appeal to some Congressional interests, is to establish basically pro-competitive policy guidelines and to grant the FCC increased jurisdiction and authority with respect to issues of market

structure and greater flexibility to forego the application of conventional public utility regulation where effective competition exists. This would resolve two major dilemmas the FCC has encountered in its attempts to introduce and maintain effective competition in the telecommunications field: first, it would remove the necessity of lengthy proceedings to prove that competition will be publicly beneficial before it can be authorized under the present statute; second, it would remove the necessity of subjecting all telecommunications carriers to the same filing, reporting, tariffing, authorization, and other procedural requirements, even though some carriers operate entirely within competitive markets where marketplace forces are demonstrably superior to regulation in satisfying user needs at just and reasonable rates. By freeing the Commission from many of its ministerial functions and due-process requirements, this option might also permit the Commission to concentrate more attention and resources on problems created by the telephone industry structure, particularly the Bell System structure. However, given the Commission's other demands both in telecommunications and in broadcasting, cable television, and radio spectrum management; its chronic and apparently worsening resource limitations; and its collegial structure, it is somewhat doubtful that the Commission could either mount an effective monitoring and enforcement program concerning Bell System operations, or itself bring about the structural changes that may be necessary if the public interest is truly to be served.

This leads to the final option, which is in effect a combination of the third option just described and the relief requested in the Department of Justice's current antitrust suit against AT&T. In addition to giving the FCC greater jurisdiction and authority concerning market structure and performance, coupled with more flexibility and a strong mandate to minimize conventional regulatory procedures and burdens, this option would segment ownership and management of the Bell System along lines comparable to those that now exist within the system itself. For example, each of the present Bell Operating Companies might be spun off and established as an independent corporation providing local exchange telephone services and related services - possibly including information-processing services - within its geographic region. AT&T's Long Lines Department, which currently owns many of the interstate long-distance facilities and manages the long-distance network and services, might be spun off together with some elements of the operating companies to form an independent supplier of long-distance telephone and other telecommunications services, operating on a fully competitive basis vis-a-vis other intercity carriers. Finally, the Western Electric Company could be spun off, perhaps split into two or more

companies, to engage in the manufacture and supply of all
forms of telecommunications and information-processing equipment, both nationally and internationally. Various arrangements could be made for the continued operation and support of
the Bell Telephone Laboratories as a major scientific resource
- for expample, as an independent organization receiving continued support from and providing assistance to both Bell operating companies and other manufacturing and service organizations, as an adjunct to the new Long Lines entity, or as a
separate public corporation.

Suggestions for restructuring or "breaking up" the Bell
System invariably generate images of cataclysmic consequences
on the part of the listener, stimulated in no small measure by
the reactions of Bell System corporate managers. Such a restructuring clearly is no minor task, and cannot be lightly
regarded. On the other hand, it would likely be far less cataclysmic, both for Bell stockholders and Bell consumers, than
one might first imagine. And in the long term, it would likely result in significant benefits to both these interested
constituencies. The only element of the System likely to be
disadvantaged by such restructuring, at least temporarily, is
the corporate management itself - which may explain at least
some of its reactions.

Conclusion

As noted previously, separation of the Bell System into
its component parts in the manner indicated would to a large
extent merely formalize operational, financial, and management
divisions that already exist within the System itself. Shareholders who previously held stock in a single company, AT&T,
would subsequently hold stock in several companies, e.g., a
local operating company, an intercity-services company, and
one or more manufacturing companies. Shareholders could subsequently decide for themselves whether to retain this multiple stock package or exchange stock so as to concentrate in-
vestments in particular companies. Since each surviving company would continue to occupy the major position in its relevant market sector, no loss in revenues or earnings should result; in fact, as a consequence of fewer regulatory constraints and more opportunities for innovation and service ex-
pansion, these companies might well be even more profitable
separately than as part of the present structure - even though
the public might also benefit from better services and lower
prices as a result of such competition. The user of basic
telephone service would probably notice few if any differences
either in the quality, cost or means of obtaining such services in the near term, although he might have more options
and improved rates in the longer term, particularly for long-
distance services. The former Bell Operating Company would

continue to provide most local-exchange telephone service, and
to handle both the set up and billing of most long-distance
calls as a result of contractual arrangements with the former
Long Lines Department. The Long Lines carrier would continue
to establish technical and operating standards for interconnection with its intercity network, which would in all probability become industry-wide standards, given the major role of
this intercity carrier. Initially, only users requiring specialized intercity services, for data transmission or other
business services, would likely require arrangements for accessing other intercity carriers. Eventually every user might
well have the option of accessing different intercity carriers
operating on a competitive basis, through the same local facilities used to provide local-exchange service.

One can hardly say this option is under active consideration in the current legislative process. It is, however, an
option which the author believes merits some consideration,
since it addresses directly the major issue that the FCC and
the Courts have confronted in virtually every major telecommunications proceeding of the past two decades: how to achieve
full, fair, and effective competition and the consumer benefits that flow therefrom, given the overwhelming dominance of
this industry by the Bell System.

Congressional subcommittees and ultimately the entire
Congress will be dealing with these issues and options over
the next few months and perhaps years. These subcommittees
seem genuinely committed to an early restatement of national
telecommunications and information policy. There are clearly
major stakes for users, new and potential competitive suppliers, and the telephone industry in this effort. The latter
group is clearly more aware, better situated, and has more resources with which to pursue its primary interests, which are
to retain control - of revenues, of operations, of technological pace. Two important sectors not effectively represented
in the debate thus far have been users and the academic community. Unfortunately, this has not been a prime subject of
academic research, nor an area of burning consumer concern, so
I do not really expect these sectors of society to have a
major direct input to the decision process.